QUANTUM PROBABILITY THEORY, PSYCHOLOGY AND LAW

This book provides the first consideration of quantum principles in legal decision making. It shows how quantum probability theory can be applied to psychology and law to deepen our understanding of psychological and legal issues, suggesting a paradigm shift in the study of these areas.

Written as an introduction to quantum probability theory for non-physicists, this book offers an accessible summary of how psychology can be applied to quantum phenomena for human decision making. It introduces psychologists and lawyers to fresh conceptual frameworks, including classical probability theory, classical statistics, data mining and quantum theory. Wojciechowski shows how quantum theory and concepts can be transferred to psychological research models and can help to resolve previously unexplained psychological phenomena, influence scientific views, undermine traditional foundations of the legal system and result in the development of new research directions. Studies conducted by the author are used to show how problems facing legal decision making can be addressed in a completely new light with the application of quantum probability theory.

This cutting-edge and innovative book will be of great interest to graduate students and researchers in the fields of forensic psychology and legal decision making, as well as legal professionals and legal scholars.

Bartosz W. Wojciechowski (PhD) is associate professor at the Jagiellonian University in Cracow, Poland; forensic psychologist; and experienced patent attorney and attorney at law. He is author and co-author of several publications on veracity assessment and application of the quantum probability theory to psychology and law.

QUANTUM PROBABILITY THEORY, PSYCHOLOGY AND LAW

Modelling Legal Decision Making with Quantum Principles

Bartosz W. Wojciechowski

Routledge
Taylor & Francis Group
LONDON AND NEW YORK

Designed cover image: © Getty

First published 2024
by Routledge
4 Park Square, Milton Park, Abingdon, Oxon OX14 4RN

and by Routledge
605 Third Avenue, New York, NY 10158

Routledge is an imprint of the Taylor & Francis Group, an informa business

British Library Cataloguing-in-Publication Data
A catalogue record for this book is available from the British Library

ISBN: 978-1-032-51487-1 (hbk)
ISBN: 978-1-032-51482-6 (pbk)
ISBN: 978-1-003-40248-0 (ebk)

DOI: 10.4324/9781003402480

Typeset in Galliard
by codeMantra

mojemu Kochanemu Kłębuszkowi

CONTENTS

ACKNOWLEDGMENTS

I would like to express my gratitude to Mrs. Elżbieta Jach, my primary school teacher and first science professor, for opening doors to the delightful and beautiful world of physics.

My special thanks go to Emmanuel M. Pothos, Paweł Błasiak and Lee C. White. Without your inspiring works, generous help and support, this book would never have been written.

My unbounded gratitude goes to Natalia, my Mother, Siggi, Grzegorz, and all family members for their continued love and care.

1

INTRODUCTION

Mathematical modelling in psychology

There are three ways to build psychological theories explaining and predicting how variables interact and relate to each other, with the use of natural verbal language, formal mathematics and computational methods (Busemeyer et al., 2015). There are numerous limitations to human intuitive and verbal reasoning, such as incapability to imagine how a dynamic system works (Hintzman, 1991), and natural language is not always suitable to describe and analyse complex processes. The intuitive reasoning and verbal reasoning of humans are severely limited. For instance, Hintzman (1991) listed at least ten significant constraints, one of which is our inability to visualize how a dynamic system operates. Formal models, including both mathematical and computer models, can circumvent these human reasoning limits.

Mathematical and computational models can address these limitations of human reasoning. Mathematical models can be used to represent quantitative relations between objects, describe governing mechanisms (laws), solve complex tasks, and evaluate and prepare forecasts (Banks & Tran, 2009; Muciek, 2012). Sometimes, the model is known prior to the experiment, and mathematical analysis is focused on the determination of the parameters, accuracy improvement and application of the model. Very often, mathematical models must be determined on the basis of analysis of the data drawn from the experiment (Muciek, 2012). Mathematics can be seen as an empirical discipline, with thoroughness not typical for any other field of research and consistent and rigorous evidence (the proof) (Suppes, 1984). One might view computational and mathematical models as a continuity of the thinking process. However, mathematical models can aid in avoiding mistakes in logic and reasoning that

DOI: 10.4324/9781003402480-1

are common in human verbal thinking (Arangala et al., 2018). The complexity of theory and data frequently necessitates the use of computers and programming languages. Almost all mathematical models can be implemented as computer models.

In the last years, remarkable growth of the computational and mathematical psychology has been observed (Busemeyer et al., 2015; Busemeyer & Bruza, 2012; Haven & Khrennikov, 2013; Sun, 2008). Theories describing and predicting how variables interact and relate to one another in psychological processes were built using natural verbal languages, and now more and more often, they are developed with the use of formal mathematics and computer approaches. Often, the verbal explanation of a psychological theory is being codified using mathematical language and then encoded into a computing language. By testing the models with actual data, the results of model fitting can give feedback for improving the models as well as our original comprehension and verbal explanations (Aerts et al., 2019; Batchelder et al., 2017; Busemeyer et al., 2015; Busemeyer & Bruza, 2012).

Mathematical models built in psychology usually can be incorporated into one of the three sets. The first category includes models that hypothesize fundamental cognitive mechanisms, such as signal detection, diffusion, information processing and reinforcement learning. They have been applied widely in a variety of formal models inside and outside psychology and cognitive research, ranging from simple visual perception to complex decision-making. The second group consists of theories that theorize basic cognitive skills such as perceptual recognition, categorization and episodic memory. The third category includes models that hypothesize higher-level cognition, such as Bayesian cognition, decision-making, semantic memory and form perception (Aerts et al., 2019; Busemeyer et al., 2015). Cognitive and decision models based on quantum probability, along neurocognitive modelling, as well as mathematical and computational modelling in clinical psychology theory, are new prospects in the area (Busemeyer & Bruza, 2012; Busemeyer et al., 2015). This book presents the application of the quantum probability theory – mathematical models developed for quantum physics, to psychology and law. We will focus on higher-level cognition and decision-making.

Classical and quantum probability theory in cognition

The modelling of cognitive processes in classical probability theory has a long history. Classical rules of logic specify, for instance, how a reasonable evaluator should alter a subjective probability estimate in light of previously gathered and new data. In traditional probability theory, occurrences are thought to be subsets of a single sample space. This indicates that the likelihood of event A can never be smaller than the probability of event A occurring in combination with event B (A and B). Nevertheless, past research has demonstrated violations of this law, known as "conjunction fallacy." The most well-known instance of this fallacy is the "Linda problem" presented by R. Tversky and D. Kahneman (1983), in which participants of a research assigned a greater likelihood that Linda is both a feminist and a bank teller than that she is merely a bank teller (probability of A and B was estimated as higher than the probability of A).

Classical models suggest that despite cognitive system's moment-to-moment change, at any given time, it is in a defined state regarding a judgement or conclusion. According to classical theories, when a judge must consider two mutually exclusive and exhaustive hypotheses – suspect is guilty or suspect is innocent – after conducting proceedings and analysing evidence, the probability assigned to guilt depends on the evidence, and at each given time, it is feasible to assign a subjective probability that either favours guilt (probability of guilt is higher than 50%) or innocence (probability of innocence is higher than 50%) or is precisely the same for both. Classical probability models presume that alternatives' values are subjectively represented prior to making a decision and that expressing a preference is the outcome of comparing these subjective representations (Busemeyer et al., 2015; White et al., 2014).

However, research shows that when the evaluation of evidence requires people to consider multiple sources of uncertainty, they make more extreme judgements than the Bayesian model dictates and tend to evaluate evidence in accordance with how coherently and comprehensively they can account for the facts (Saks & Thompson, 2003). A substantial amount of empirical evidence supports the notion that decisions can have a constructive influence (Ariely & Norton, 2008; Kahneman & Snell, 1992) and that the act of selecting affects later preferences for the available options and modifies their relative merits. Studies indicate that selecting a particular option increases our preference for it (Ariely & Norton, 2008; Kahneman & Snell, 1992); affective evaluation for positively and negatively valenced stimuli is constructive (White et al., 2014, 2016); and simple evaluation of one well-known person's trustworthiness alters affective judgement about other individuals (Wang et al., 2014; White et al., 2016). Participants do not maintain the likelihood of an occurrence or the utility of an outcome constant across contexts, despite the fact that the logic dictates that they should (Narens, 2014). Human judgements are frequently constructed from the cognitive state for the question at hand, and drawing a conclusion from one question changes the context and disturbs the cognitive system, which in turn affects the answer to the next question, producing order effects (Bruza et al., 2015; Wang et al., 2014). When a memory cue for a displayed or unpresented item is administered on a memory test, the object has the potential to simultaneously occupy all members of a mutually exclusive set of episodic states, but once the cue is administered, it occupies only one of those states (Brainerd et al., 2013). Prior research reveals that classical probability principles are broken in both basic cognitive tasks and more complex decision-making processes (Pothos & Trueblood, 2015).

Previous psychological research has shown that human judgement frequently differs from traditional normative standards and that under uncertainty, behaviour can be irrational; there are numerous well-established empirical findings that are difficult to reconcile with classical probability principles (Kahneman, 2012; Tversky & Kahneman, 1974, 1986). In a variety of reasoning tasks, people frequently provide incorrect answers according to logic or probability theory (DeNeys & Goel, 2011). In transparent situations, the axioms of rational choice are often met, but they are frequently broken in opaque ones (Tversky & Kahneman, 1986). It has been established that people do not act in accordance with the assumptions of the expected utility theory (Kahneman, 2012); instead, they adhere to the sure-thing principle; choosing one option appears to influence subsequent preferences for the options available, resulting in order effects (White et al., 2016); and actions do not merely reveal preferences but create them (Ariely & Norton, 2008). Yet, human judgements are frequently not simply retrieved from memory, but rather created from the cognitive state of the topic at hand (Wang et al., 2014).

In the early 20th century, scientists realized that classical physics did not adequately explain the outcomes of some experiments. Classical physics delivered appropriate outcomes for systems far bigger than atoms but failed for small-scale systems. Quantum mechanics is the study of the behaviour of matter and energy at the sub-nuclear, molecule, atomic and nuclear levels. It is a mathematical instrument used to anticipate the behaviour of elementary particles. It is the most scientifically supported scientific hypothesis in human history (Wang et al., 2013). Quantum physics is regarded as humanity's greatest successful scientific theoretical breakthrough and one of the most fundamental contributions to the creation of contemporary civilization. It was developed to explain observations that were incomprehensible under the previous classical physical theory. The quantum theory introduced a brand-new set of revolutionary concepts.

Quantum physics is considered to be the most successful scientific theoretical achievement that humans have ever created and one of the most significant for the development of modern civilization. It has been created to explain findings that were impossible to understand using the older classical physical theory. Quantum theory brought an entirely new set of revolutionary principles. Quantum mechanics is sometimes regarded as the most flawless physical theory ever formulated, and its prediction accuracy is remarkable. For example, in quantum electrodynamics (QED), which describes the interaction of light and matter (i.e. electrons and photons), the agreement of theoretical predictions with the results of measurements of a certain physical quantity, called the magnetic moment of the electron, is on the order of ten to the 11th power, which Richard P. Feynman likened to measuring the distance from Los Angeles to New York (approximately 4,400 kilometres) with a ruler containing (about 0.05 mm). The practical applications of quantum mechanics, on the contrary, are completely ubiquitous: from computers through nanotechnology and, in the near future, quantum computers. Furthermore, quantum mechanics provides the only explanation for seemingly simple phenomena, such as why, when sitting on a chair, we do not fall through its surface and down due to the Earth's gravitational force, despite the fact that our bodies (and chairs) are composed of 99.99% vacuum (Łukasik, 2017).

The biggest achievements of the physical sciences are based on extremely general quantitative laws that serve as the basis for understanding a vast array of specific physical systems (Chater & Brown, 2008). To what extent is it possible to develop broad principles in the cognitive sciences that may be utilized to describe specific aspects of perception, memory and decision-making, based on quantum theory? Quantum cognition, a relatively new field of study that employs concepts and methods from quantum theory and quantum probability, appears to be an especially useful and promising framework for describing a wide range of seemingly irrational judgement and decision-making findings, thus providing a new perspective for cognitive science (Khrennikov, 2010). The fundamental hypothesis is that human thought, under a number of crucial bounded-rationality circumstances, adheres to quantum probability laws as opposed to classical probability principles. Quantum cognition permits the consideration of perplexing effects like conjunction fallacies, disjunction fallacies, averaging effects, unpacking effects, order effects, violations of the sure-thing principle in decision theory, violations of symmetry in similarity judgements and the generation of novel predictions like the quantum question (QQ) equality (Aerts et al., 2019; Busemeyer et al., 2015; Wang & Busemeyer, 2013).

Quantum cognition study applies the mathematically abstract ideas of quantum theory to cognitive science concerns (Wang et al., 2013). In quantum probability theory, probabilistic judgement is frequently very reliant on context and order. Individual states can be superposition states (the state has not yet been determined, all possible values in the superposition have the potential to be expressed, and it is impossible to assign a specific value prior to measurement) and composite systems can be entangled (properties of one element depend on the state of its twin, entangled element; they cannot be decomposed into their subsystems; psychological measurements must be made sequentially; and context must be taken into account) (Busemeyer & Wang, 2015).

Classical probability theory is founded on a different set of assumptions than quantum probability theory, which defines how probabilities should be given to events. It is based on linear algebra and a variety of additional assumptions and theorems. Representation in quantum probability theory is built on a multidimensional space, Hilbert space, in which distinct subspaces correspond to distinct things. In this space, the current state of the system is defined by a vector; a projection "lays down" a vector on a certain subspace. According to one of the fundamental theorems of quantum probability, the squared length of a projection along a subspace indicates the chance that the associated possibility, that the state vector reflects the system, is true (Pothos et al., 2013; White et al., 2014). By projecting the state onto distinct subspaces and calculating the square length of the projected vector, the consistency between the state vector and these other entities in quantum space may be determined (Busemeyer et al., 2015; Pothos et al., 2013).

Quantum probability theory offers an alternative explanation for the conjunction fallacy to the classical probability theory ("the Linda problem"). Referring to the quantum entanglement principle permits an explanation of the order effect in terms of context-dependent probability assessment. It is considered that a joint probability distribution cannot be specified from the probability distributions of its individual parts. Qualities such as being a bank teller and a feminist must be evaluated sequentially, as the context provided by the initial assessment disrupts subsequent evaluations (Pothos & Busemeyer, 2013). The "Linda problem" is addressed by a series of projections of state vectors onto a two-dimensional plane that correspond to the "bank teller" and "feminist" possibilities. Various subspaces represent various cognitive state vector possibilities. The higher the relationship between two incompatible outcomes, the narrower the angle between their respective subspaces. A small angle between state vectors can be equated to a high correlation coefficient. When there is a small angle, a sequential projection of the state vector from one subspace to the other loses minimal amplitude; hence, accepting one conclusion makes the other outcome very probable (Pothos & Busemeyer, 2013).

In conventional cognitive models, it is thought that a person's judgement and cognition are in a fixed condition at any given time. Research on cognitive system's sensitivity to measurements, order effects of measurement, conflict, ambiguity, uncertainty and violations of classical probability laws in cognitive and choice studies demonstrate that this assumption is false (Wang et al., 2013). The superposition principle has been valuable in cognitive research and has helped to explain puzzling results that were incomprehensible using typical classical probability models (Busemeyer & Wang, 2015; Pothos et al., 2015). In the superposition state, it is thought that the conclusion, result or measurement is inconsistent with any single conceivable consequence. To be in a state of superposition indicates that all conceivable measurement values have the potential to be expressed at each moment and that there is merely the possibility of different outcomes; if the cognitive system evolves with time, so does the potential for each possibility (Busemeyer & Wang, 2015; Pothos & Busemeyer, 2013).

Quantum probability theory concepts are applicable not only to physics but also to any subject in which formalization of uncertainty is required, and scholars are researching its applications in fields as diverse as economics and information theory (Aerts et al., 2019; Gallus et al., 2023; Haven & Khrennikov, 2013; Khrennikov, 2014, 2010; Pothos, Busemeyer, 2013). In its most rudimentary form, quantum cognition is merely an algorithm for computing probabilities, with no claim to reflect the fundamental manner in which the brain makes decisions (Yearsley & Busemeyer, 2016). With this style of thinking, the success or failure of a method is solely determined by the empirical accuracy of its forecasts. Yet, attempts are being made to regard quantum cognitive theories as process models that do in some way reflect the process of arriving at a particular judgement (Kvam et al., 2015). Contextuality, interference effects, order effects and constructive judgements, to mention a few, appear to correlate nicely with how we believe humans receive information.

The pervasiveness of the heuristic system and the fact that human reasoning often depends on intuitive heuristics rather than a purposeful, controlled reasoning process explain violations of established normative criteria. Neurobiologists and neuropsychologists assert that there are two distinct human reasoning systems, dual brain pathways: one for heuristic thinking and one for analytical reasoning (DeNeys & Goel, 2011). In the past decade, the number of research examining the probable relevance of physical–mathematical multidimensional or quantum notions to comprehending the human mind in a psychological and neurological setting has increased significantly. It is anticipated that quantum and multidimensional methods will aid in describing and modelling both brain and mental activities (Korf, 2015). According to a number of authors (Bruza & Busemeyer, 2012; Korf, 2015; Pothos & Busemeyer, 2013), higher mental activities may be described by physical–mathematical quantum models (Bruza & Busemeyer, 2012; Korf, 2015; Pothos & Busemeyer, 2013), and quantum models are recognized as outperforming traditional models when it comes to fitting human data and predicting their behaviour (Bruza et al., 2015; Busemeyer et al., 2015; Pothos & Busemeyer, 2009; Pothos et al., 2015; Wang et al., 2014). Quantum theory has been effective in describing decision-making, which looks troublesome under conventional formalism (Pothos et al., 2013) and shows promise in tackling cognitive processes that have proven difficult to explain using classical probability theory (Bruza et al., 2015). Quantum concepts provide obvious and clear explanations for these discoveries.

Quantum cognition is an emerging field, but it has already been applied to studies on bistable perception, overdistribution in episodic memory, entanglement in associative memory, violations of rational decision-making, probability judgement errors, over- and under-extension in conceptual combinations, order effects on inference, causal reasoning, asymmetric similarity judgements and vagueness (Busemeyer & Wang, 2015; Busemeyer et al., 2015; Pothos & Busemeyer, 2022, 2009, 2013; Pothos & Trueblood, 2015). However, quantum models of decision-making do not typically imply that the brain's fundamental functioning is represented by quantum mechanics. In this way, they differ from so-called quantum brain models (Hameroff & Penrose, 1996). It is important to note that the quantum approach to cognitive modelling makes no assumptions about neural implementation nor quantum neural operations.

Rationale for quantum probability theory in psychology and law

There are numerous reasons to employ the quantum theory to human cognition and many arguments (Wang et al., 2013). According to Busemeyer and Bruza (2012), there are six reasons for the application of the quantum approach to cognition and decision. First is that quantum theory allows a person to be in a superposition state. An indefinite or superposition state means that, for example, member of a jury cannot be assumed to be either in the state of conviction that the suspect is definitely guilty or in the state of conviction that the suspect is not guilty at some point, both possibilities have the potential (called state amplitudes) to be expressed at any one time (Heisenberg, 1958). This of course does not imply that the jury member is physically present in both states at the same time but means that the psychological state is suspended somewhere in between, not yet determined. Intuitively, if the decision-maker is in an indeterminate state, does not necessarily believe the person is guilty nor does not necessarily believe the person is innocent. Instead, there is a superposition state that leaves the jury member conflicted, ambiguous, perplexed or unsure regarding the suspect's guilt.

At a given moment, the potential for guilt may be greater than the potential for innocence, and these potentials (amplitudes) may shift from one moment to the next, but both solutions are possibly available at any moment. In quantum theory, there is no single trajectory or sample path over time prior to making a determination. Rather, there is a flow of potentials across states, across time. In this way, quantum theory enables one to model the cognitive system as a wave travelling across state space in time until a decision is made. Nonetheless, once a decision is made and uncertainty is eliminated, the state becomes definitive, as if the wave had collapsed into a particle-like point. Consequently, quantum systems necessitate both wave (uncertain) and particle (certain) perspectives of a cognitive system. We can assume that the wave character of an indefinite state reflects the psychological experience of conflict, ambiguity, confusion and uncertainty, whereas the particle nature of a definite state captures the psychological experience of conflict resolution, choice and certainty.

Second is that judgements produce rather than document. According to numerous formal models, the cognitive system may change from moment to moment, but the information we record at a given time reflects the system's state as it existed just prior to our inquiry. Formal cognitive models presume, for instance, that a person's response to a question such as "*Are you sad?*" reflects his or her state just before to the question being posed. If we instead ask the person, "*Are you happy?*", their response would again represent their state shortly before we pose the first question. Taking a measurement of a system generates rather than records a property of the system. This is one of the more perplexing conclusions learnt from quantum theory (Peres, 1980).

Before posing a question, a quantum system may exist in an indeterminate state. A person may be unsure of his or her emotions, for instance, after viewing a touching movie scene. Quantum systems produce a response based on the interaction between the system's indefinite state and the query posed (Bohr, 1958). This interaction transforms an undefined state into a definite one. For instance, the person's feelings following the movie scene may have been equivocal, but this becomes more clear after answering the question regarding mood. If the response is "*Yes, I am sad*" the individual responds accordingly. In fact, this is the foundation for contemporary psychological theories of emotion. Additionally, decision scientists think that beliefs and preferences are created as opposed to just being read out from memory (Payne et al., 1993). For instance, a person may initially be indecisive about a collection of artworks on show, but if requested to select one as a gift, a preference order is established for the purpose. We do not aim to argue that every response to every inquiry necessitates the formation of an opinion. For some queries, we have a saved response that may be retrieved on demand. Other questions, however, are novel and more difficult, requiring the respondent to formulate an answer based on current condition and circumstances. Therefore, Busemeyer and Bruza (2012) argue that the quantum principle of generating a reality from an interaction between the person's indeterminate state and the posed question matches psychological intuition for complicated judgements better than the assumption that the answer merely reflects a predetermined state.

The third reason is associated with the influence of subsequent judgements. According to quantum theory, if a person begins in an indeterminate state and is then asked a question, the response to the question will shift the state to one that is more definitive in relation to the question asked. However, this shift in state after the first question prompts a distinct response to the following questions, making the order of inquiry significant. We can consider a following example, we can ask a witness: "*How confident are you, that you have seen the suspect at the crime scene?*" the response we would expect is, "*I am absolutely certain.*" However, if the senior witness is first asked, "*How large is your visual impairment?*" this witness typically responds, "*I have two pairs of glasses, but can still watch TV.*" Following this sobering response, a subsequent inquiry regarding the level of confidence elicits a less optimistic response. Therefore, the first question establishes a context that modifies the response to the second question. Hence, we cannot define a combined probability of responses to questions A and B; rather, we can only assign a probability to the sequence of responses to questions A followed by B. If A and B are two measurements in quantum physics, and the probabilities of the results depend on the sequence of the experiments, then A and B are non-commutative. In physics, measurements of position and momentum along the same direction are non-commutative, whereas measurements of position along horizontal and vertical coordinates are commutative. Developing a probabilistic model for non-commutative measurements gives rise to many of the mathematical features of quantum theory, including Heisenberg's famous uncertainty principle (Heisenberg, 1958). Additionally, order effects introduce uncertainty into a person's judgements. If the response to the first question A provides a definite state with regard to that question, the state established by A may be indefinite with regard to the second question B. A definite state (technically, an *eigenstate*) for one question is an indefinite state (superposition) for another. If the issues are deemed incompatible, the mathematical implementation of this incompatibility is the non-commutativity of quantum measurements. Question order effects are a significant concern for attitude researchers, who seek a quantum-like theoretical understanding of these effects (Feldman & Lynch, 1988).

The fourth reason is that judgements do not always adhere to conventional logic. The classical probability theory utilised in contemporary cognition and decision models is derived from the Kolmogorov axioms (Kolmogorov, 1933). These axioms assign probability to sets representing events. Consequently, the Kolmogorov theory's family of sets adheres to the Boolean axioms of logic and serves as the cornerstone for Bayes net inferences. Research shows, however, that the law of total probability is broken by the outcomes of the disjunction experiment, the category – decision-making experiment, and the two-slit experiments. Von Neumann's (1932) axioms are the basis for the quantum probability theory. These axioms apply probabilities to events specified as vector space subspaces. The definite states serve as the foundation for the vector space, and an indefinite or superposition state can be any point inside this vector space. The logic of subspaces does not adhere to the distributive axiom of Boolean logic, which is a significant consequence of their use (Hughes, 1989) and allows more ambiguous thoughts represented by a superposition of qualities that are either true or false. Since quantum logic does not always respect the distributive axiom, the quantum model does not always obey the law of total probability (Khrennikov, 2010). The quantum model can therefore explain the outcomes of the disjunction experiment in psychology and the two-slit experiment in physics. Busemeyer and Bruza (2012) believe that the traditional logic and probability theory are too restrictive to describe human decisions and judgements. Quantum logic can therefore be perceived as a generalisation of classical logic, whereas quantum probability can be perceived as a generalization of probability theory.

Several studies demonstrate that judgements contradict the concept of uniqueness. Classical probability theory (Kolmogorov, 1933) used in modern cognitive and decision models is based on the concept of uniqueness (Griffiths et al., 2010). A single sample space that describes exhaustively and comprehensively all potential experimental events is assumed. This follows from the classical Boolean algebra: if A is an event and B is another event from an experiment, then A and B must also be an event, and repeated application of this principle results in intersections that cannot be further subdivided (the atoms or elements or points of the sample space). Unions of atoms, elements or sample space points can be used to define every occurrence. Given considerable thinking, this is a significant constraint on any theory. According to Busemeyer and Bruza (2012), it oversimplifies the tremendously complex structure of our reality. Assuming uniqueness in probability judgement tests involving humans has consequences. We ask a participant to describe the probabilities of future political occurrences in an experiment. Perhaps a single sample space can do this. However, we can ask the same person to describe the probability of future scientific advancement occurrences. It's hard to assume that one person can assign probabilities to all historical and scientific events. The individual may have to rely on one description of occurrences (one sample space) for political prospects and another for future scientific developments. Uniqueness is not assumed in quantum probability. Incompatible questions cannot be analysed using the same criteria, and hence, distinct sample spaces must be constructed. Quantum theory allows the use of a partial Boolean algebra: one sample space can be used to answer a first set of questions in a Boolean manner, and another sample space can be used to answer a second set. Both Boolean subalgebras are pasted together in a coherent but non-Boolean manner. This provides for greater flexibility in assigning probability to occurrences and does not require the development of all conceivable joint probabilities, which is assumed to be necessary to understand the complexity of human cognition and decision-making.

The sixth reason according to Busemeyer and Bruza (2012) is that cognitive phenomena may not be decomposable. In cognitive science, researchers frequently describe a cognitive process for a task by providing a vast collection of random variables, although only a tiny sample of these variables is seen under any given experimental condition. Then, it is assumed that there exists a single complete joint distribution, which is a joint probability distribution across all random variables, that can be used to calculate the observed marginalization distributions for any subset of variables. When attempting to physically demonstrate the existence of quantum entanglement, quantum scientists questioned this very basic and innocuous theoretical approach. Entanglement is a remarkable phenomenon in which two ostensibly separate and distinct systems behave as if they were one – these systems are also referred to as "quantum linked." It was discovered that it is impossible to design such a comprehensive joint distribution when systems are entangled. This finding intuitively suggests that there is an extreme form of dependencies between the systems beyond those derived from conventional probability theory. Since the original probabilistic models for entanglement experiments were created, there has been a substantial body of literature discussing the probabilistic basis of quantum correlations. This has produced analysis approaches that are sufficiently generic to be applied to cognitive science. A word of warning, however:

it is highly speculative to assume the existence of quantum supercorrelations in cognition. There is, however, a continuously expanding body of groundbreaking literature supporting this viewpoint. But what are the implications of quantum correlations for our cognitive models? Reductionism has served as an influential philosophical foundation for model scientific creation. This refers to the concept that phenomena can be investigated by separating their many components and then combining the resulting results. Decomposable phenomena or systems are phenomena or systems that can be understood in this manner. Non-decomposable systems cannot be comprehended so easily. Most of the cognitive science models are decomposable, as their fundamental pieces and their relationships are understood. Consider, for instance, the spreading activation model of human memory for words. Even though the network structure may have a high degree of interconnectedness, the model nevertheless considers the nodes (i.e., the words) to be separate components. This is typical of many cognitive science paradigms. But we can ask if reductionism is always suitable for cognitive modelling? The presence of quantum correlations in cognitive phenomena suggests that this assumption can at least be questioned in certain circumstances. For instance, when a word is studied in a memory experiment, it is believed that its associative network emerges in tandem with the researched word. This interpretation is based on an intuition very similar to quantum entanglement: the study word and its associations behave as one. Existence of quantum correlations shows that the cognitive model in question is not decomposable in the manner we initially anticipated and specified using a certain set of random variables. It compels us to radically reconsider the nature of the phenomenon being modelled.

Quantum physics was created to explain puzzling facts that could not be understood using traditional classical physical concepts. Physicists were obliged to adopt fundamentally new ways of thinking in order to establish quantum mechanics, resulting in a revolutionary understanding of probability. Currently, the cognitive sciences are seeing a similar change. In the past, classical probability theory and mathematical concepts of classical mechanics served as the foundation for the vast bulk of cognitive modelling research. Although substantial progress has been made in comprehending cognition in this fashion, empirical evidence has accumulated that appears unusual in the context of standard probability theory. To explain all sorts of decision errors and prejudice, for example, multiple heuristic toolboxes had to be developed to circumvent or supplement the conventional theoretical framework in decision research (Gigerenzer & Goldstein, 1996; Payne et al., 1993; Tversky & Kahneman, 1974). Quantum theory, in contrast, provides a coherent and systematic framework for explaining such surprising facts (e.g., Busemeyer & Bruza, 2012; Busemeyer et al., 2011).

Quantum theory can specifically provide new light on five persisting challenges in cognitive modelling. The difficulty of precisely describing the psychological concepts of conflict, ambiguity and unpredictability is the first obstacle. Traditional cognitive models assume that a person is in a definite (technically, non-dispersive) state with respect to certain judgement and cognition at each moment; however, the true state of the person at each moment is unknown to the modeller; therefore, the model can only assign a probability to a cognitive response with a value at each moment. This type of model is stochastic because the modeller is unsure of the trajectory (i.e., the definite state at each time point) that an individual is following. It is stated as if the cognitive system's state follows a well-defined trajectory in its state space. A quantum account, on the contrary, allows a person to live in an indeterminate (technically, dispersive) condition known as a superposition state at any one time. This strictly implies that one cannot presume that psychological states are characterized by certain values that can be documented by a psychological assessment at every instant. Being in a state of superposition denotes that all possible definite values within the superposition are capable of being articulated at every instant (Heisenberg, 1958). A superposition state represents the conflict, ambiguity or uncertainty perceived by cognitive processes (Blutner et al., 2013; Brainerd et al., 2013; Wang & Busemeyer, 2013). Quantum modelling enables us to specify the state of a cognitive system as it progresses through time in its state space (Atmanspacher et al., 2004; Busemeyer et al., 2006) until a decision is reached, at which point the state collapses to a definite value.

Conventional cognitive models assume that what we measure and record at a given time reflects the state of the cognitive system immediately preceding assessment. For instance, the response to a judgement question just reflects the state of the question just before to our posing it. However, it is challenging to formalize the measurement sensitivity of the system following the classical probability theory. Whereas one of the more intriguing ideas derived from quantum theory is that measurement of a system not only records a property of the system but also establishes the state (Peres, 1998). Just before a question is posed, a quantum system can exist in a state of superposition. The response we obtain from the system is the outcome of the interplay between its condition and the query asked (Bohr, 1958). This interaction transforms a superposition state into a definite or clearly defined state. Consistent with the constructive view in psychological research (e.g., Feldman & Lynch, 1988; Schwarz, 2007) and the idea that choice can alter preference (e.g., Ariely & Norton, 2008; Sharot et al., 2010) is the quantum principle of constructing a reality from an interaction between an individual's indeterminate state and the question being asked. In fact, it fits more closely with psychological intuition for complex judgements, decisions and other cognitive evaluations than the assumption that the measured response is only a register of a pre-existing condition or returns a pre-existing cognitive property.

There is also the challenge of formalizing the order effects of cognitive tests. The change in cognitive processes that results from answering one question can influence how one responds to future questions (Moore, 2002). In other words, the context of subsequent cognitive tests is modified by the original cognitive test, and the order of cognitive tests becomes relevant. Similar order effects have been observed and characterized by psychological and public opinion researchers (e.g., Moore, 2002). In order to develop formal cognitive models, we cannot describe the joint probability of responses to a series of questions. Instead, we can only assign the answer sequence a probability. In quantum physics, order-dependent measurements are non-commutative operations (Wang & Busemeyer, 2013). Quantum theory derives many of its mathematical components, including Heisenberg's uncertainty principle, from the creation of a probabilistic model for non-commutative observations (Heisenberg, 1958). Measurement order effects, such as question order effects, are a major concern in attitude, judgement and decision research, which seeks a theoretical understanding of the order effects similar to that attained in quantum theory (e.g., Feldman & Lynch, 1988; Wang & Busemeyer, 2013).

Human mind and judgement do not always adhere to the fundamental rules of logic and probability (for a summary, see Busemeyer & Bruza, 2012; Busemeyer et al., 2011). The classical probabilities applied in modern cognition and decision models are derived from the Kolmogorov axioms (1933) that assign probabilities to sets of events. The set family of the Kolmogorov theory adheres to the Boolean axioms of logic. The distributive axiom is among the most essential axioms in Boolean logic. The law of total probability can be inferred from this distributive axiom, so setting the framework for Bayes net conclusions. Notwithstanding this, a number of psychological tests reveal that the law of total probability is broken. The axioms of Von Neumann (1955) allocate probabilities to events specified as vector space subspaces. From these assumptions derive the quantum probability theory. Hughes (1989) found that representing events by subspaces requires a logic of subspaces and does not comply with the distributive axiom of Boolean logic (despite though they constitute a partial Boolean algebra structure). This indicates quantum models do not always adhere to the law of total probability (Busemeyer & Wang, 2017; Khrennikov, 2010; Pothos & Busemeyer, 2009). Quantum probability is a generalization of probability theory, whereas quantum logic is a generalization of classical logic. Aerts and Durt (1994) argue that classical probability theory may be insufficient to explain human cognition and decision-making, as indicated by the growing empirical evidence that contradicts classical probability principles.

In cognitive science, it is commonly believed that concepts and processes are "decomposable" such that the whole can be understood in terms of its constituent parts. This decomposability is reflected by the common assumption that a complete joint probability distribution exists across all cognitive system measures, from which it is possible to reconstruct a pairwise joint distribution for any pair of observations. Nonetheless, the psychological literature gives ample evidence to the contrary. Facial cognition, for example, is characterized by a heavy dependence on the holistic qualities of faces (e.g., Farah et al., 1998; Wenger & Townsend, 2001), and it has been proposed that the perception of ambiguous figures is a holistic process (e.g., Atmanspacher et al., 2004; Long & Toppino, 2004). Combinations of concepts cannot be broken down into meaningful components (Aerts & Durt, 1994; Aerts & Sozzo, 2011; Blutner et al., 2013; Bruza et al., 2009). In contrast to classical models, quantum models enable us to represent cognitive systems as non-decomposable; therefore, it is possible that pairwise probabilities cannot be derived from a shared joint probability distribution. This demonstrates the existence of quantum theory's entanglement correlations, which are significantly different from those expected by classical models.

Quantum probability theory in research on legal decision-making

At the beginning of the 20th century, physicists were forced to develop a new theory. The principles of quantum mechanics have been proposed to explain findings in physics that seemed paradoxical from a classical point of view. Previous assumptions did not allow to explain results of experiments indicating that measurement is not a simple read out from a pre-existing or recorded state; instead, it is constructed from the current context and act of measuring; making a measurement changes the context which disturbs the physical system; and changes in context produced by the first measurement affect the next measurement producing order effects, so that physical measurement does not obey the commutative rule of classic probability theory.

It may seem surprising or even bizarre to draw a connection between physics, psychology and law research. However, many studies conducted previously and those conducted by the author, presented in the book, confirm that quantum probability theory not only is inspiring but also allows to deepen our understanding of the psychological and legal issues. There is a growing number of researchers examining quantum theory as a way to understand perplexing findings and problems. Physicist, mathematicians and cognitivists are conducting many studies on the application of the quantum probability theory to cognition. Some aspects of quantum theory, such as contextuality and entanglement, open the door to address problems facing cognition and decision in a completely new light.

So far, the application of the quantum probability theory has been focused mostly on data from human experimental paradigms (Pothos & Yearsley, 2023). Such datasets are inevitably limited and subject to all sort of biases, including task demands and the more likely engagement of fast cognition. The putative added value of the quantum cognitive models can be best illustrated with applications employing decision-making data from the real world. Works of few authors point in this direction (e.g. Gallus et al., 2023; Khrennikov, 2010; White et al., 2014; Wojciechowski & Pothos, 2018; Wojciechowski et al., 2022), but quantum models are being tested mostly and continually in experimental studies, in labs or using surveys circulated to paid participants. We share a firm belief, expressed by Pothos and Yearsley (2023), that there is much to be gained by a greater focus of the quantum cognition studies on real-world data.

The criminal trial is an example of a situation in which judgements are made under ambiguous situations. At the earliest phases of recognition, the determining authorities are unaware of the investigation's real course of events. The court does not have a strong understanding of which hypotheses are most probable, does not know which hypotheses are most likely to be true, and lacks knowledge of both what has occurred and what may occur after the trial. The purpose of the evidence gathered during the proceedings is to enable the court to create an exhaustive and accurate catalogue of hypotheses about the state of affairs, i.e., to enable a comprehensive and in-depth analysis of all legal and factual issues pertaining to the individual's behaviour under consideration – the alleged act. In addition, the conclusions drawn from the content of the evidence make it possible to estimate the value of possible outcomes of the action and facilitate the estimation of the probability that the hypotheses regarding guilt and criminal responsibility formulated by the trial decision-maker are accurate.

The judicial system is built on the premise that judges are capable of assessing evidence logically and reaching accurate decisions. Thus, it is essential to comprehend the mechanisms and study the cognitive consequences and errors that can interfere with the criminal responsibility decision-making process (Simon, 2012). Prior research by forensic psychologists has primarily focused on the impact of external factors on the judicial process, such as the content of expert opinions, media attention, the use of inadmissible evidence, stereotypes and biases, the witness's confidence in his own testimony, or the physical attractiveness of either the suspect, the persons being questioned or prosecuted (Daftar-Kapur et al., 2010; Fox et al., 2011; Kapardis, 2003; McAuliff et al, 2003; McCabe & Krauss, 2011; Spellman & Tenney, 2010). Previous research has demonstrated that judges are occasionally biased and prone to judgement errors, have trouble correctly calculating probabilities, lack an understanding of statistical principles and make diverse decisions based on the same evidence (Green & Wrightsman, 2003; Spellman & Tenney, 2010). The findings also imply that the way in which Bayes model is used by lawyers leads to violations of the laws of classical probability theory. Judges compelled to assess and evaluate various sources of information about a crime, while making choices under conditions of ambiguity, tend to voice more extreme opinions and make more radical trial verdicts than classical probability models would suggest (Saks & Thompson, 2003). Classical probability theories fail to explain the mechanisms guiding individual behaviour in instances of ambiguity or uncertainty, which is why they are sometimes characterized as irrational or illogical. Nonetheless, the same processes can be precisely defined and explained using mathematical models from quantum physics (Bruza & Busemeyer, 2012; Pothos & Busemeyer, 2009, 2013).

Trueblood and Busemeyer (2011) asked provocatively – what if we substitute physical measurement with human judgement and physical system with cognitive system? And the aim of this book is to prompt researchers to reflect on the following question – what if we use quantum probability theory to analyse legal decision-making? The purpose of this book is to introduce to psychologists and lawyers fresh conceptual frameworks, based in quantum theory, and to show the application of quantum theory to law and psychology.

Quantum decision-making offers a new pool of hypotheses regarding decisions of interest to the law. For instance, Jaeger and Trueblood (2019) identified eight examples of legal predictions that arise from the assumptions of quantum probability theory. According to the authors, the final judgements of the factfinders can be influenced by quantum effects, such as quantum Zeno effect; a systematic bias in responses of the eyewitnesses answering sequentially presented line-ups can be explained by a quantum prediction that sequential assessment (as opposed to simultaneous line-up) can be biased; quantum theory predicts that the order of items on the ballot will influence results; verdicts in cases involving multiple charges (or claims) or multiple defendants will be influenced by order effects, similarly to quantum context and order effects. Jaeger and Trueblood (2019) also indicate that based on quantum predictions, we can expect that factfinders at trial will not weigh each piece of evidence independently or make judgements independently, but decision-maker's beliefs are shaped by previously considered facts and finings, influencing the assessment of each piece of evidence that follows, potentially having disproportionate influence of the relevant judgement; legal conclusions will be affected by the order in which courts consider relevant authorities; preliminary decisions and categorizations of the law enforcement officers may affect their actions (categorization before engaging suspects may result in lower likelihood of responding with force); and last, but not least, if heuristics and biases are treated as products of people's use of incompatible mental representations, it is possible to debias decision-makers by getting them to use compatible representations.

The subject matter of this book

The purpose of this book is to introduce to psychologists and lawyers fresh conceptual frameworks, based in quantum theory, and to show the application of quantum theory to law and psychology. This book describes the author's scientific journey, the transition from classical probability theory, classical statistics and data mining to quantum theory. This book presents differences between the classical and quantum probability theories and introduces chosen quantum phenomena to psychology and law research. Quantum theory and concepts are transferred to psychological research models. Results of the studies conducted by the author are used to show how problems facing legal decision-making can be addressed in a completely new light, with the application of quantum probability theory.

This book is an introduction to quantum probability theory for non-physicists and summary of psychological studies on the application of quantum phenomena to human decision-making. It can influence future studies on legal decision-making, as well as cause a paradigm shift in psychology and law. The primary goal of this book is to show how quantum concepts can be used to model human cognition in legal settings. Previous research showed the application of quantum theory in judgement, conceptual combinations, memory and perception (Busemeyer et al., 2015). Several studies showing that cognitive processes in legal contexts can be modelled and interpreted using quantum models are described. We will present the results of research focused on hypotheses such as:

- when making legal decisions judges, prosecutors, attorneys and others are in the superposition state;
- quantum entanglement principle applies to legal decision-making;
- quantum theory provides a coherent explanation for order effect and conjunction fallacy present in the legal decision-making.

Most of the previous studies were introductory and focused on general (cognitive) psychology, rather than applied (legal) psychology. Books and papers published to date very often are written by physicists or mathematicians. In these publications, quantum theory is described with equations and forms, which are difficult to understand for scholars without mathematical background. Very often, authors of the books and papers are focusing on the analysis of various approaches to quantum theory, development of formal models and focus on theoretical disputes (convincing about the superiority of one model over another). Jennifer Trueblood is one of few researchers and mathematical psychologists publishing papers and chapters on quantum decision-making and law-related decisions (Jaeger & Trueblood, 2019). However, they are thematically limited to issues such as juror decision-making, witness line-ups and policing. This book will focus on different problems of psychology and law.

The second chapter will present the basic assumptions of the quantum probability theory. Both the classical and the quantum probability theories will be described and confronted. The rules for quantum probabilities differ from those of classical probabilities. In quantum cognitive modelling, states of individual quantum systems are called pure states, which are represented as vectors in a vector space (the Hilbert space). Properties of the Hilbert space and vectors' operations will be explained. Two aspects of quantum theory open the door to addressing problems facing cognition and decision in a totally new light. The first is known as "contextuality" of judgements and decisions: the context generated by making a first judgement or decision interferes with subsequent judgements or decisions to produce order effects, so that judgements and decisions are non-commutative. The second aspect relates to quantum entanglement – a phenomenon whereby making an observation on one part of the system instantaneously affects the state in another part of the system, even if the respective systems are separated by space-like distances. This opens the door to developing quantum-like models of cognitive phenomena which are not decompositional in nature. Examples of cognitive processes modelled using the quantum account and results of the previous studies will be presented.

In the third chapter, we will focus on superposition and contextuality. To be in a superposition state means that all possible definite values within the superposition have the potential for being expressed at each moment. A superposition state provides an intrinsic representation of the conflict, ambiguity or uncertainty that people experience in cognitive processes. In this chapter, we will present results of research conducted on a group of 60 lawyers (judges, prosecutors and attorneys at law) and 30 lay persons aimed at seeking answers to the following questions: what type of probability theory best describes the way lawyers make judgements under uncertainty and decision under conflict? Can their decisions be explained and predicted using a quantum theory? Could quantum probability theory provide us with any advantages in cognitive modelling? Are judges in a superposition, when they decide on the guilt or innocence of suspects?

In the fourth chapter, we will focus on quantum entanglement and order effects. We will be modelling the influence of the intermediate judgement and conjunction fallacy with quantum probability principles with the use of quantum model to derive qualitative predictions for conjunction errors devised by Busemeyer and Bruza (2012) and quantum cognitive model formalizing constructive evaluations proposed by White and colleagues (White et al., 2014, 2016). Quantum entanglement is a phenomenon whereby making an observation on one part of the system instantaneously affects the state in another part of the system, even if the respective systems are separated by space-like distances. Quantum probability theory predicts that when judges rule on multiple offences, a difference in the probability of suspect's guilt judgement should be observed. Probability of conviction will vary depending on whether suspect's guilt of the first and second offence will be rated individually or if it will be rated in conjunction. It is assumed that after rating the probability of suspect's guilt, revision in the cognitive state occurs. The original state vector changes to a new conditional state vector, which is the projection onto the subspace representing "yes" or "no" answer to the question of guilt.

Fifth chapter will introduce the quantum Zeno effect and help to establish if this quantum phenomenon can be observed in criminal procedure. The quantum Zeno effect is the suppression of unitary time evolution in quantum systems and can be caused by measurement or interactions with the environment. It is a feature of systems under which time evolution of a particle can be slowed down by measuring it frequently enough. In other words, when a system is being constantly watched, it cannot change. Evolution of the system can be slowed down or even frozen by the act of frequent measurement or interaction with the environment. In this chapter, we will present the results of a study conducted on a group of 45 judges adjudicating in criminal cases. We will analyse if preliminary rulings, interim orders and other numerous intermediate decisions related to the course of the criminal trial affect the final judgement of the case.

In the sixth chapter, we will study the process of a statement's veracity assessment and seek answers to the question of whether those evaluating testimony or explanations are in a state of superposition. Content analysis methods are used to assess the veracity of witnesses' and suspects' statements. They are based on the Undeutsch hypothesis assuming that there are quantitative and qualitative differences between statements derived from the memory of real-life experiences and fabricated or fictitious accounts. We will present the results of the research conducted on a group of 40 specialists trained in content analysis. The aim of the research was to establish if raters are in a superposition when assessing the veracity of a statement given in criminal cases. Traditional cognitive models assume that what we measure and record at a particular moment reflects the state of the cognitive system as it existed immediately before we measure it. For example, the answer to a judgement question simply reflects the state regarding this question just before we asked it. One of the more provocative lessons learnt from quantum theory is that a measurement of a system both records and creates a property of the system. Immediately before asking a question, a quantum system can be in a superposition state. The answer we obtain from the system is constructed from the interaction of its state and the question that we ask. This interaction creates a definite or sharply defined state out of a superposition state.

The last chapter will present the results of an experiment aimed at determining if the level of conformity of the prisoners' and suspects' answers brakes the theoretical limits set by the Bell's inequality. We treat a pair of individuals as a system, with each individual as a subsystem. These two individuals had some interaction, then they go their separate ways. For example, the individuals could be two suspects awaiting trial or two prisoners planning their escape from prison. First, they make their plan (interact), and then, they separate, as they put the plan into action and go their different ways. Then, we can ask those individuals pairs of binary questions, for example, concerning some aspect of the escape plan. We already know that prisoners interacted with each other, prior to their separation. So, we already anticipate that there would be some correlation in the answers they provide. Assuming a classical theoretical picture, there is a common cause in the answers provided by prisoners, which can lead to correlations in the answers. However, the size of these correlations needs to be bounded by a limit expressed in Bell's inequalities.

References

Aerts, D., & Durt, T. (1994). Quantum, classical and intermediate: An illustrative example. *Foundations of Physics*, Vol. 23, pp. 1353–1369; DOI: 10.1007/BF02283037.

Aerts, D., Khrennikov, A., Melucci, M., & Toni, B. (2019). *Quantum – Like Models for Information Retrieval and Decision – Making*. Springer, Cham.

Aerts, D., & Sozzo, S. (2011). *Quantum Structure in Cognition: Why and How Concepts Are Entangled*. In Proceedings of the Quantum Interaction Conference. Springer.

Arangala, C., Luke, N.S., & Yokley, K.A. (2018). *Mathematical Modeling. Branching Beyond Calculus*. CRC Press, Boca Raton.

Ariely, D., & Norton, M.I. (2008). How actions create – not just reveal – preferences. *Trends in Cognitive Sciences*, Vol. 12, pp. 13–16; DOI: 10.1016/j.tics.2007.10.008.

Atmanspacher, H., Filk, T., & Roemer, H. (2004). Quantum Zeno features of bistable perception. *Biological Cybernetics*, Vol. 90, pp. 33–40; DOI: 10.1007/s00422-003-0436-4.

Banks, H.T., & Tran, H.T. (2009). *Mathematical and Experimental Modeling of Physical and Biological Processes*. CRC Press, Boca Raton.

Batchelder, W.H., Colonius, H., Dzhafarov, E.N., & Myung, J. (2017). *New Handbook of Mathematical Psychology, Vol. 1: Foundations and Methodology*. Cambridge University Press, Cambridge.

Blutner, R., Pothos, E.M., & Bruza, P. (2013). A quantum probability perspective on borderline vagueness. *Topics in Cognitive Sciences*, Vol. 5, pp. 711–736; DOI: 10.1111/tops.12041.

Bohr, N. (1958). *Atomic Physics and Human Knowledge*. Wiley and Sons, New York.

Brainerd, Ch.J., Wang, Z., & Reyna, V.F. (2013). Superposition of episodic memories: Overdistribution and quantum models. *Topics in Cognitive Sciences*, Vol. 5, pp. 773–799; DOI: 10.1111/tops.12039.

Bruza, P.D., & Busemeyer, J.R. (2012). *Quantum Models of Cognition and Decision*. Cambridge University Press, Cambridge.

Bruza, P.D., Kitto, K., Nelson, D., & McEvoy, C. (2009). Is there something quantum – lkike about the human mental lexicon? *Journal of Mathematical Psychology*, Vol. 53, No. 5, pp. 362–377; DOI: 10.1016/j.jmp.2009.04.004.

Bruza, P. D., Kitto, K., Ramm, B., & Sitbon, L. (2015). A probabilistic framework for analysing the compositionality of conceptual combinations. *Journal of Mathematical Psychology*, Vol. 67, pp. 26–38.

Bruza, P.D., Wang, Z.J., & Busemeyer, J.R. (2015). Quantum cognition: A new theoretical approach to psychology. *Trends in Cognitive Sciences*, Vol. 19, No. 7, pp. 383–393; DOI: 10.1016/j.tics.2015.05.001.

Busemeyer, J.R., & Bruza, P.D. (2012). *Quantum Models of Cognition and Decision.* Cambridge University Press, Cambridge.

Busemeyer, J.R., Pothos, E.M., Franco, R., & Trueblood, J.S. (2011). A quantum theoretical explanation for probability judgment errors. *Psychological Review,* Vol. 118, No. 2, pp. 193–218; DOI: 10.1037/a0022542.

Busemeyer, J.R., & Wang, Z. (2015). What is quantum cognition, and how is it applied to psychology? *Current Directions in Psychological Science,* Vol. 24, No. 3, pp. 163–169; DOI: 10.1177/0963721414568663.

Busemeyer, J.R., & Wang, Z. (2017). Is there a problem with quantum models of psychological measurements? *PLoS ONE,* Vol. 12, No. 11; DOI: 10.1371/journal.pone.0187733.

Busemeyer, J.R., Wang, Z., & Pothos, E. (2015). Quantum models of cognition and decision, [in:] *Oxford Handbook of Computational and Mathematical Psychology,* [eds.] Busemeyer, J.R., Wang, Z., Townsend, J.T., & Eidels, A., Oxford University Press, New York, pp. 369–389.

Busemeyer, J.R., Wang, Z., Pothos, E., & Trueblood, J.S. (2015). The conjunction fallacy, confirmation and quantum theory: Comment on Tentori, Crupi and Russo (2013). *Journal of Experimental Psychology: General,* Vol. 144, No. 1, pp. 236–243; DOI: 10.1037/xge0000035.

Busemeyer J. R., Wang, Z., & Townsend, J. T. (2006). Quantum dynamics of human decision-making. *Journal of Mathematical Psychology,* Vol. 50, pp. 220–241; DOI: 10.1016/j.jmp.2006.01.003.

Busemeyer, J.R., Wang, Z., Townsend, J.T., & Eidels, A. (2015). *The Oxford Handbook of Computational and Mathematical Psychology.* Oxford University Press, Oxford.

Chater, N., & Brown, G.D.A. (2008). From universal laws of cognition to specific cognitive models. *Cognitive Science,* Vol. 32, pp. 36–67; DOI: 10.1080/0364021 0701801941.

Daftary-Kapur, T., Dumas, R., & Penrod, S.D. (2010). Jury decision-making biases and methods to counter them. *Legal and Criminological Psychology,* Vol. 15, pp. 133–154; DOI: 10.1348/135532509X465624.

DeNeys, W., & Goel, V. (2011). Heuristics and biases in the brain: Dual neural pathways for decision making, [in:] *Neuroscience of Decision Making,* [eds.] Vartanian, O., & Mandel, D.R., Psychology Press, New York, pp. 125–141.

Farah, M.J., Wilson, K.D., Drain, M., & Tanaka, J.N. (1998). What is "special about face perception? *Psychological Review,* Vol. 105, No. 3, pp. 482–498; DOI: 10.1037/0033–295X.105.3.482.

Feldman, J.M., & Lynch, J.G. (1988). Self – generated validity and other effects of measurement on belief, attitude, intention, and behaviour. *Journal of Applied Psychology,* Vol. 73, No. 3, pp. 421–435; DOI: 10.1037/0021–9010.73.3.421.

Fox, P., Wingrove, T., & Pfeifer, C. (2011). A comparison of students' and Jury Panelists' decision-making in split recovery cases. *Behavioral Sciences and the Law,* Vol. 29, pp. 358–375; DOI: 10.1002/bsl.989.

Gallus, C., Pothos, E.M., Blasiak, P., Yearsley, J.M., & Wojciechowski, B.W. (2023). Bell correlations outside physics. *Scientific Reports,* Vol. 13, No. 4394; DOI: 10.1038/s41598-023-31441-x.

Gigerenzer, G., & Goldstein, D. (1996). Reasoning the fast and frugal way: Models of bounded rationality. *Psychological Review,* Vol. 103, pp. 650–669; DOI: 10.1037/0033–295x.103.4.650.

Green, E., & Wrightsman, L. (2003). Decision making by juries and judges: International perspective, [in:] *Handbook of Psychology in Legal Contexts,* [eds.] Carson, D., & Bull, R., John Wiley and Sons, Chichester, pp. 401–422.

Griffiths, T.L., Chater, N., Kemp, C., Perfors, A., & Tenenbaum, J. B. (2010). Probabilistic models of cognition: exploring representations and inductive biases. *Trends in Cognitive Sciences*, Vol. 14, pp. 357–364; DOI: 10.1016/j.tics.2010.05.004.

Hameroff, S., & Penrose, R. (1996). Orchestrated reduction of quantum coherence in brain microtubules: A model for consciousness. *Neural Network World*, Vol. 5, pp. 793–804; DOI: 10.1016/0378-4754(96)80476-9.

Haven, E., & Khrennikov, A. (2013). *Quantum Social Science*. Cambridge University Press, Cambridge.

Heisenberg, W. (1958). *Physics and Philosophy: The Revolution in Modern Science, Role of Modern Physics in Human Thinking*. Harper & Row, New York.

Hintzman, D.L. (1991). Why are formal models useful in psychology? [in:] *Relating Theory and Data: Essays on Human Memory in Honor of Bennet B. Murdock*, [eds.] Hockley, W. E., & Lewandowsky, S., Erlbaum, Hillsdale, NJ, pp. 39–56.

Hughes, R.I.G. (1989). *The Structure and Interpretation of Quantum Mechanics*. Harvard University Press, Cambridge, MA.

Jaeger, Ch.B., & Trueblood, J.S. (2019). Thinking quantum: A new perspective on decisionmaking in law. *Florida State University Law Review*, Vol. 46, No. 4, pp. 733–806.

Kahneman, D. (2012). *Pułapki Myślenia. O Myśleniu Szybkim I Wolnym*. Media Rodzina, Poznań.

Kahneman, D., & Snell, J. (1992). Predicting a changing taste: Do people know what they will like? *Journal of Behavioral Decision Making*, Vol. 5, No. 3, pp. 187–200; DOI: 10.1002/bdm.3960050304.

Kapardis, A. (2003). *Psychology and Law. A Critical Introduction*. Cambridge University Press, Cambridge.

Khrennikov, A. (2010). *Ubiquitious Quantum Structure*. From Psychology to Finance. Springer, Heidelberg.

Khrennikov, A. (2014). *Beyond Quantum*. Taylor & Francis Group, New York.

Kolmogorov, A.N. (1933). *Foundations of Probability Theory*. Julius Springer, Berlin.

Korf, J. (2015). Quantum and multidimensional explanations in a neurobiological context of mind. *The Neuroscientist*, Vol. 21, No. 4, pp. 345–355; DOI: 10.1177/1073858414542250.

Kvam, P.D., Pleskac, T.J., Yu, S., & Busemeyer, J.R. (2015). Interference effects of choice on confidence: Quantum characteristics of evidence accumulation. *PNAS*, Vol. 112, No. 34, pp. 10645–10650; DOI: 10.1073/pnas.1500688112.

Long, G.M., & Toppino, T.C. (2004). Enduring interest in perceptual ambiguity: Alternating views of reversible figures. *Psychological Bulletin*, Vol. 130, No. 5, pp. 748–768; DOI: 10.1037/0033-2909.130.5.748.

Łukasik, A. (2017). *Filozoficzne Zagadnienia Mechaniki Kwantowej*. Wydawnictwo UMCS, Lublin.

McAuliff, B.D., Nemeth, R.J., Bornstein, B.H., & Penrod, S.D. (2003). Juror decision - making in the twenty - first century: Confronting science and technology in court, [in:] *Handbook of Psychology in Legal Contexts*, [eds.] Carson, D., & Bull, R., John Wiley and Sons, Chichester, pp. 303–372.

McCabe, J.G., & Krauss, D.A. (2011). The effect of acknowledging Mock Jurors' feelings on affective and cognitive biases: It depends on the sample. *Behavioral Sciences and the Law*, Vol. 29, pp. 331–357; DOI: 10.1002/bsl.990.

Moore, D.W. (2002). Measuring new types of question-order effects. *Public Opinion Quarterly*, Vol. 66, No. 1, pp. 80–91; DOI: 10.1086/338631.

Muciek, A. (2012). *Wyznaczanie Modeli Matematycznych z Danych Eksperymentalnych*. Oficyna Wydawnicza Politechniki Wrocławskiej, Wrocław.

Narens, L. (2014). Alternative probability theories for cognitive psychology. *Topics in Cognitive Science*, Vol. 6, pp. 114–120; DOI: 10.1111/tops.12071.

Payne, J.W., Bettman, J.R., & Johnson, J. (1993). *The Adaptive Decision Maker.* Cambridge University Press, Cambridge.

Peres, A. (1980). Zeno paradox in quantum theory. *American Journal of Physics*, Vol. 48, pp. 931–932; DOI: 10.1119/1.12204.

Pothos, E.M., Barque-Duran, A., Yearsley, J.M., Trueblood, J.S., Busemeyer, J.R., & Hampton J.A. (2015). Progress and current challenges with the quantum similarity model. *Frontiers in Psychology*, Vol. 6, No. 205; DOI: 10.3389/fpsyg.2015.00205

Pothos, E.M., & Busemeyer, J.R. (2009). A quantum probability explanation for violations of "rational" decision theory. *Proceedings of the Royal Society B: Biological Sciences*, Vol. 276, No. 1665, pp. 2171–2178; DOI: 10.1098/rspb.2009.0121.

Pothos, E.M., & Busemeyer, J.R. (2013). Can quantum probability provide a new direction for cognitive modeling? *Behavioral and Brain Sciences*, Vol. 36, pp. 255–327; DOI: 10.1017/S0140525X12001525.

Pothos, E.M., Busemeyer, J.R., & Trueblood, J.S. (2013). A quantum geometric model of similarity. *Psychological Review*, Vol. 120, No. 3, pp. 679–696; DOI: 10.1037/a0033142.

Pothos, E.M., & Busemeyer, J.R. (2022). Quantum cognition. *Annual Reviews in Psychology*, Vol. 73, pp. 749–778; DOI: 10.1146/annurev-psych-033020–123501.

Pothos, E.M., Shiffrin, R.M., & Busemeyer, J.R. (2013). The dynamics of decision making when probabilities are vaguely specified. *Journal of Mathematical Psychology*, Vol. 59, pp. 6–17; DOI: 10.1016/j.jmp.2013.09.001.

Pothos, E.M., & Trueblood, J.S. (2015). Structured representations in a quantum probability model of similarity. *Journal of Mathematical Psychology*, Vol. 64–65, pp. 35–43; DOI: 10.1016/j.jmp.2014.12.001.

Pothos, E.M., & Yearsley, J.M. (2023). Quantum cognition: Quo vadis? [in:] *The Quantum – Like Revolution,* [eds.] Plotnitsky, A., & Haven, E., Springer Nature, Cham, pp 281–297; DOI:10.1007/978-3-031-12986-5_14.

Saks, M.J., & Thompson, W.C. (2003). Assessing evidence: Proving facts, [in:] *Handbook of Psychology in Legal Contexts*, [eds.] Carson, D., & Bull, R., John Wiley and Sons, Chichester, pp. 329–345.

Schwarz, N. (2007). Attitude construction: Evaluation in context. *Social Cognition*, Vol. 25, No. 5, pp. 638–656; DOI: 10.1521/soco.2007.25.5.638

Sharot, T., Velasquez, C.M., & Dolan, R.J. (2010). Do decisions shape preference? : Evidence from blind choice, *Psychological Science*, Vol. 21, No. 9. pp. 1231–1235; DOI: 10.1177/0956797610379235.

Simon. D, (2012). *In doubt. The Psychology of the Criminal Justice Process.* Harvard University Press, Cambridge, MA, London.

Spellman, B.A., & Tenney, E.R. (2010). Credible testimony in and out of court. *Psychonomic Bulletin & Review*, Vol. 17, No. 2, pp. 168–173; DOI: 10.3758/PBR.17.2.168.

Sun, R. (2008). *The Cambridge Handbook of Computational Psychology.* Cambridge University Press, Cambridge.

Suppes, P. (1984). Conflicting Intuitions about causality. *Midwest Studies in Philosophy*, Vol. 9, No. 1, pp. 151–168; DOI: 10.1111/j.1475–4975.1984.tb00057.x

Trueblood, J.S., & Busemeyer, J.R. (2011a). A quantum probability account of order effects in inference. *Cognitive Science*, Vol. 11, pp. 1518–1552; DOI: 10.1111/j.1551–6709.2011.01197.x.

Trueblood, J.S., & Busemeyer, J.R. (2011b). A quantum probability model of casual reasoning. *Frontiers in Cognitive Science*, Vol. 3; DOI: 10.3389/fp-syg.2012.00138.
Tversky, R., & Kahneman, D. (1974). Judgment under uncertainty: Heuristics and biases. *Science*, Vol. 185, pp. 1124–1130.
Tversky, R., & Kahneman, D. (1983). Extensional versus intuitive reasoning: The conjunctive fallacy in probability judgment. *Psychological Review*, Vol. 90, pp. 293–315.
Tversky, R., & Kahneman, D. (1986). Rational choice and the framing of decisions. *The Journal of Business*, Vol. 59, No. 4, Part 2: The Behavioral Foundations of Economic Theory, pp. S251–S278.
von Neumann, J. (1955). *Mathematical Foundations of Quantum Mechanics*. Princeton University Press, Princeton.
Wang, Z., & Busemeyer, J.R. (2013). A quantum question order model supported by empirical tests of an a priori and precise prediction. *Topics in Cognitive* Science, Vol. 5, No. 4; Doi: 10.1111/tops.12040.
Wang, Z., Busemeyer, J.R., Atmanspacher, H., & Pothos, E.M. (2013). The potential of using quantum theory to build models of cognition. *Topics in Cognitive Sciences*, Vol. 5, pp. 672–688; DOI: 10.1111/tops.12043.
Wang, Z., Solloway, T., Shiffrin, R.M., & Busemeyer, J.R. (2014). Context effects produced by questions orders reveal quantum nature of human judgment. *PNAS*, Vol. 111, No. 26, pp. 9431–9436; DOI: 10.1073/pnas.1407756111
Wenger, M.J., & Townsend, J.T. [eds.] (2001). *Computational, Geometric, and Process Perspectives on Facial Cognition: Contexts and Challenges*. Lawrence Erlbaum Associates Publishers, Mahwah.
White, L.C., Barque – Duran, A., & Pothos, E.M. (2016). An investigation of a quantum probability model for the constructive effect of affective evaluation. *Philosophical Transactions A*, Vol. 374, No. 20150142; DOI: 10.1098/rsta.2015.0142.
White, L.C., Pothos, E.M., & Busemeyer, J.R. (2014). Sometimes it does hurt to ask: The constructive role of articulating impression. *Cognition*, Vol. 133, pp. 48–64; DOI: 10.1016/j.cognition.2014.05.015.
Wojciechowski, B.W., Izydoryczk, B., Blasiak, P., Yearsley, J., White, L.C., & Pothos, E.M. (2022). Constructive biases in clinical judgment. *Topics in Cognitive Science*, Vol. 14, pp. 508–527; DOI: 10.1111/tops.12547.
Wojciechowski, B.W., & Pothos, E.M. (2018). Is there a conjunction fallacy in legal probabilistic making? *Frontiers in Psychology*, Vol. 9, No. 391; DOI: 10.3389/fpsyg.2018.00391.
Yearsley, J.M., & Busemeyer, J.R. (2016). Quantum cognition and decision theories: A tutorial. *Journal of Mathematical Psychology*, Vol. 74, pp. 99–116; DOI: 10.1016/j.jmp.2015.11.005.

2

BASIC ASSUMPTIONS OF THE QUANTUM COGNITION

Axioms of the classical probability theory

Assumptions of the classical probability theory remain at the heart of traditional scientific theories. They are so commonly and widely applied and accepted, that we take them for granted and presume them to be obviously true (Busemeyer & Wang, 2015). Classical probability theory evolved over several centuries, beginning in the 18th century with contributions by Pascal and Laplace. According to Laplace's classical definition of probability, the likelihood of an event is equal to the ratio of the number of events favouring its occurrence to all potential events, if an event can happen in one of mutually exclusive and equally plausible ways. Probability is a real number between 0 and 1. However, an axiomatic foundation for classical probability theory did not exist until Kolmogorov (1933) provided one. The Kolmogorov probability calculus has been the accepted theory of probability in mathematics and science since the 1930s. It is regarded by many statisticians and philosophers of science as the cornerstone of a comprehensive, logical theory of belief incorporating uncertainty. Much of the theory was initially motivated by problems arising in physics, and later applications appeared in economics, engineering, insurance, statistics and in many more disciplines.

Classical probability theory is founded on the premise that elementary events are represented as subsets of a larger set called the sample space. Similar to the concepts of point, line and plane in geometry, the idea of a random event is regarded here as a basic concept which is not defined. When two or more elementary events occur (events A and B), they are treated as complex events and can lead to: the conjunction of events A and B, which can be observed when both A and B occur; the alternative of events A or B, which

DOI: 10.4324/9781003402480-2

occurs when either A or B occur, or both occurrences can occur concurrently. The product of events A and B can be represented as the product of the sets B and A (intersection of both A and B), the alternative of A and B (A or B) to the sum of the sets A and B (union), and the opposite event to A, that is not-A, is the complementary part of the set A. Events that cannot occur simultaneously are called mutually exclusive events. An example would be tossing a coin – head and tail are events which cannot occur at the same time. An event that never occurs is called an impossible event. An event that always occurs is called a certain event. As the independent events are regarded those for which the occurrence of one does not affect the occurrence of the other. Events are called dependent events, when one has effect on the other.

According to Kolmogorov axioms (1933), probability of an event is a function of the events space and is an additive measure, its value ranges from 0 to 1, and for certain events, it is equal to 1 (Narens, 2007). If events A and B are independent (there is no intersection of subsets), probability of the sum of both events is equal to sum of probability of an event A and probability of an event B (complement); if A and B are mutually exclusive and complementing, probability of the sum of events A and B is equal to sum of probabilities of event A and event B. If A and B are two events, it is always possible to define product of the events A and B which is equal to product of probability of event A and probability of event B, given that event A occurred, which is equal to product of probability of event B and probability of event A, given that event B occurred. It is worth noting, that the order of events A and B does not matter, and in both cases product reflecting probability of an event will be the same. If event A is independent from probability of event B, product of the probabilities of events is equal to product of probability of event A and probability of event B. According to the classical formula for total probability, if event A occurs, it can occur in two mutually exclusive and exhaustive ways: either A occurs together with event B, or A occurs without event B (that is, with an event opposite to B) (Kolmogorov, 1933; Łukasik, 2017; Narens, 2007).

The assumptions of the classical probability theory presented so far presumably will be welcomed and accepted by most readers. Classical probability theory has a long history of modelling cognitive processes and classical assumptions remain at the heart of traditional cognitive theories (Busemeyer & Wang, 2018). Social and behavioural scientists, as well as lawyers are generally trained to accept these axioms (explicitly or implicitly), and consequently most of us consider the theory as the only way to think about events and probabilities.

The axioms of the classical probability theory are a straightforward collection of rules (e.g., Howson & Urbach, 1993). Whether probabilities be evaluated as frequencies or subjective degrees of belief, the same rules apply. If probabilities are interpreted as frequencies, one obtains a representation of probabilities based on parts of generalized volumes and is lead to the axioms of classical probability theory. Psychologically, the normative case of classical probability theory is supported by significant mathematical conclusions, such as the Dutch book theorem, which states that classically consistent probabilistic assignment protects against certain loss (de Finetti, 1993; Pothos et al., 2017). Using the Dutch book theorem, these same axioms are derived when subjective degrees of belief are considered (de Finetti et al., 1993). In other words, the classical probability framework can be derived from either frequentist probabilities (via set theory) or subjective probabilities (with the aid of the Dutch book theorem). Classical probability theory is applicable whenever an individual must change subjective probabilities in light of new or existing evidence. Some excellent cognitive models based on classical probability theory provide solid evidence that humans are predominantly rational (Oaksford & Chater, 2009).

There have been persistent discrepancies between the predictions of classical probability theory and human behaviour, and these findings cast doubt on any important claims on human rationality in decision-making. Such outcomes are frequently referred to as fallacies to indicate that the corresponding decision-making is flawed. The conjunction fallacy, identified by Tversky and Kahneman (1983) in one of the most prominent studies on decision-making, is one of the most well-known fallacies. Tversky and Kahneman provided participants with brief descriptions of fictitious individuals. In one of their cases, Linda was shown as a smart and opinionated individual. After reading the narrative, participants were asked to rank a series of propositions about Linda according to their perceived likelihood. The essential statements were that Linda is a feminist, a bank teller and both. Tversky and Kahneman's findings suggested that according to the participants of the study probability of Linda being a bank teller and a feminist was higher than probability of Linda being just a bank teller. This finding is known as the critical finding since it is not possible in classical probability theory. The conjunction fallacy has been thoroughly repeated, in both Tversky and Kahneman's original study and later works. The conjunction fallacy is obviously not prevalent in decision-making, but it is straightforward to establish conditions that reliably lead to the fallacy (Busemeyer et al., 2011). Many attempts have been made to reconcile the conjunction fallacy with the classical probability theory principles and rational decision-making. Influential reasoning holds that the conjunction fallacy occurs due to a misunderstanding between the experimenter's intentions and the participant's comprehension. Despite the fact that various disambiguation techniques can lessen the frequency of conjunction fallacy, residual conjunction fallacies persist (Dulany & Hilton, 1991; Moro, 2009; Sides et al., 2002; Tentori & Crupi, 2012).

Why are the finding of Tversky and Kahneman so troublesome for the evaluation of human rationality in decision-making? As previously said, classical probability theory implies a notion of probabilistic assignment as generalized volumes, with subsets of events corresponding to portions of volumes – a conjunction is the intersection of two conjuncts, and it can never be greater than either conjunct. Many attempts have been made to exonerate the conjunction of fallacy, even discounting concerns over how participants interpret the key claims. When participants evaluate the conjunction, they are in a different mental state or context than when they evaluate the conjunct. Instead, it is common to suppose that there are two distinct decision-making processes: one that is fully normative and yields conclusions according with the principles of classical probability theory, and one that is rapid and approximative, for quick, somewhat insignificant decisions (DeNeys & Goel, 2011; Elqayam & Evans, 2013; Kahneman, 2001; Sloman, 1996).

Moreover, there are some critical, hidden assumptions upon which traditional cognitive models are based (Busemeyer & Wang, 2015). Classical probability theory obeys the commutative axiom. According to this assumption, the order of two propositions does not matter when they are considered. We can consider two sentences – "*a defendant is guilty*" and "*a defendant should be sentenced to imprisonment.*" Both sentences can be accepted or rejected. Following the commutative axiom, we can consider either probability of concluding that a defendant is guilty and deciding to punish the defendant or we can conclude that the defendant should be sentenced and then consider the probability that he or she is guilty. Moreover, it would be possible to rule on punishment, without even assessing probability of suspect's guilt. This set of questions can be considered in different orders, but following the assumptions of classical probability theory should lead to the same outcome. Classical logic is also based on the distributive axiom, which can be formulated as follows: punishment should be concluded either the defendant is guilty, or the defendant is not guilty. In other words, the likelihood that a penalty is selected should equal the probability that the defendant is deemed guilty and a punishment is selected, plus the probability that the defendant is deemed innocent and a punishment is selected (Busemeyer & Wang, 2018).

Assumptions of the quantum probability theory

Empirical findings that are troublesome from a classical perspective, such as the conjunction fallacy, have motivated the pursuit of cognitive models based on the assumptions of the quantum probability theory (QPT). They have been used to accommodate a variety of choice errors, such as order effects in evidence evaluation and legal decision-making (Aerts & Aerts, 1995; Busemeyer et al., 2011; Busemeyer & Bruza, 2012; Haven & Khrennikov, 2013; Pothos & Busemeyer, 2009, 2013; Trueblood & Busemeyer, 2011; Wang et al., 2014; White et al., 2014, 2016).

QPT is a formal framework for assessing probability to events, abstract foundation of quantum physics, but not specifically tied to physics (Atmanspacher et al., 2002; Busemeyer et al., 2015; Khrennikov, 2010; Pothos & Busemeyer, 2013). It is applicable not just in physics, but in any science in which there is a need to formalize uncertainty. QPT has been applied in areas as diverse as economics, information theory, biology and finance (Aerts et al., 2019; Khrennikov, 1997, 2010, 2013, 2014; Pothos & Busemeyer, 2013). Superposition, entanglement, incompatibility and interference are aspects of QPT that make it a promising framework for understanding cognition (Pothos & Busemeyer, 2013). It is worth to mention that the idea of using quantum theory in psychology has existed for nearly 100 years. Niels Bohr, one of the founding fathers of quantum mechanics, was known to believe that some aspects of quantum theory could provide insight about cognitive processes. Moreover, Edgar Rubin, a Danish psychologist, acquainted Bohr with the idea of complementarity coined by William James, and the concept of complementarity of incompatible observables has become one of the foundations of quantum mechanics (Busemeyer et al., 2015).

Incompatibility is one of the crucial characteristics of QPT that differentiates it from the classical probability theory. There is a fundamental distinction between compatible and incompatible questions for cognitive system (in classical probability theory all questions are assumed by default to be compatible). Whether some questions are considered compatible, or incompatible is part of the analysis that specifies the corresponding cognitive model. In the psychological terms, incompatibility between questions means, that a cognitive agent cannot formulate a single thought for combination of the corresponding outcomes (Pothos & Busemeyer, 2013). Psychological measurements (i.e., judgements) often require one to take different perspectives, which have to be taken sequentially and context generated by the first measure disturbs subsequent measurements (Busemeyer & Wang, 2015). If two questions A and B are incompatible, it is not possible to define a single question regarding their conjunction. Answer to question A causes a superposition state regarding question B – if A is true, then B can neither be true nor false at the same time. In the QPT conjunction of the incompatible questions is defined in a sequential way – question A and then B, or question B and then question A. The answer to question A can affect the consideration of question B (and vice versa), so it is predicted that answers for both questions can be different depending on the order of asking. This novel way to think of probability is key to some of the most puzzling predictions of quantum physics, but it also allows modelling of the cognitive interference and order effects (Pothos & Busemeyer, 2013).

In the classical probability theory, uncertainty concerns lack of knowledge about the state of the system that determines answers to the questions. In the QPT, uncertainty has a deeper notion, and it arises when a cognitive system is in a superposition among different possible outcomes. To be in a superposition state means that all possible measurement values have some potential for being expressed at each given moment. Superposition means that a given moment state is not consistent with any single possible outcome. The potentials can interfere with each other (similar to how waves interfere) leading to the change of the value of the final measurement (Busemeyer & Wang, 2015). We can only estimate potentiality of different possible outcomes, and if the cognitive system evolves in time – the probability for each possibility is evolving as well. A court may rule in favour of punishment while remaining superposed (i.e., unresolved or uncertain) about the issue of guilt. Quantum probabilities allow violation of the classical law of total probability – the probability of ruling whether to impose a punishment, while remaining in superposition with respect to guilt, may differ from the total probability of deciding to punish, after resolving the question of guilt. Superposition reflects intuitive way in which we perceive fuzziness, conflict, ambiguity and ambivalence (Busemeyer & Wang, 2015; Pothos & Busemeyer, 2013). Furthermore, quantum probabilities are non-additive and transition between states depend decisively on the prior measurement.

Compositionality of complex cognitive systems can be represented by entanglement. When there is an entangled composite system, it is not possible to limit change of one element of the system, without affecting the other part. QPT allows the specification of a composite system of entangled elements for which it is not possible to specify any joint probability distribution from the probability distribution of the constituent parts. This can allow to model interdependencies among the constituent parts not possible in classical probability theory, and startling predictions, especially when the parts of the entangled system are spatially or temporally separated.

The components of any quantum cognitive model include a space of possible thoughts, judgements or outcomes with subspaces corresponding to particular beliefs, opinions and choices, a cognitive state that keeps track of a decision-maker's current mental state, and a dynamics that specifies how this mental state changes with time and evidence presentation. We can compute probabilities for any relevant judgement outcome based on these variables (Yearsley & Busemeyer, 2016). The starting point in a quantum model is a vector space (called Hilbert space), where all possible outcomes for questions (observables and propositions) are represented as subspaces. In quantum theory the Hilbert spaces we consider are complex spaces, which means they can be spanned by vectors whose entries are complex numbers (Yearsley & Busemeyer, 2016). All possible outcomes are represented as subspaces of this vector space, each possible question or event can be represented in a different dimension of the vector space. The overall dimensionality of the vector space (number of dimensions and subspaces) is determined by the number of possible unique elements or states of the analysed system. In the simplest case, these subspaces are one dimensional. The second element is a normalized vector, which has the role to represent the mental state just prior to answering questions. The state vector embodies all our current knowledge of the cognitive system under consideration. In QPT, probabilistic assessment is always state (mental) dependent, while in classical probability theory (if the mental state is taken into consideration at all), it has to be concluded as conditionalizing information. Any vector can be expressed as a linear combination of the vectors, so that they form a basis for the multidimensional space. Amplitudes reflect components of the state vector along the different basis vector.

For example, if we consider a hypothetical question of whether suspect is guilty of two crimes – crime A, and crime B, one dimensional space (called ray) in the vector space would correspond to the probability that suspect committed crime A, and the second space would represent the probability of committing crime B. To compute probabilities with consideration of the QPT rules – the probability that judge will answer "yes" to the question regarding suspect's guilt is represented by the projection onto the subspace representing the "yes" outcome to the question. A projector operator takes a vector and lays it down onto subspace. Probability that a given person has a particular property depends upon the projection onto the subspace corresponding to that property (Busemeyer et al., 2015; Pothos & Busemeyer, 2013; Yearsley & Busemeyer, 2016). Probability of the answer "yes" is equal to the squared length of vector's projection. Figure 2.1 illustrates a two-dimensional Hilbert space and the quantum-like representation of a legal decision-making process.

The mental state of the legal decision-maker can be represented by a normalized (to the length of one unit) vector, denoted as ψ. There is a set of rays for the possibility that the observer is in a purely suspect guilty vs suspect innocent cognitive state (result of the case summary assessment) and likewise there is a set of rays for the combined cognitive impact of processing the exonerating or incriminating description. Ray proximity indicates consistency for the corresponding possibilities. Probability for a question outcome – answer to the question regarding suspect's guilt is the squared length of the projection of the mental state vector, to the subspace corresponding to the question outcome. For example, to compute the probability that the judge finds the suspect guilty of committing crime A, we project (lay down) the vector ψ onto the crime A subspace and compute the square of the length of this projection. So, the higher overlap between the mental state vector and a given subspace corresponding to crime A or crime B indicates higher probability of finding the defendant guilty.

What happens if the judge considers the evidence and then decides that the suspect has committed a crime? In quantum theory, the mental state vector has to identify with the judgement outcome, so the new mental state vector has to be a normalized vector along the guilty ray, denoted as ψ. This is a very constrained prediction for what happens to the mental state following a judgement (and concerns a fundamental theorem in quantum theory). Outside quantum theory there is no reason why the judgement could not produce a mental state that is halfway between the judgement outcome and the original

FIGURE 2.1 An illustration of the quantum probability approach for the single conjunction fallacy in assessment of probability that suspect committed two crimes.

state or indeed leave the original mental state unaffected. It is worth to note, that in case of incompatible questions (such as – is suspect guilty of crime A, is suspect guilty of crime B) conjunction for the questions can only be computed as a sequential projection, since we cannot concurrently determine the truth or falsity of incompatible questions.

We can predict that the end result from a sequential projection following two questions (is the suspect guilty of crime A and is the suspect guilty of crime B) will be different in comparison to projection of the answer to the first question or to the answer to the second question, but asked as the first one or in separation. Decision about suspect's guilt leads to a collapse of the state vector on the first subspace, the state vector changes position in such a way, as to align with the question outcome. When the decision-maker resolves question of defendant's guilt, the new state will be aligned with the "guilty" ray; resolving the first question changes the judge's mental state (Pothos & Busemeyer, 2022). Because the second question is answered from a different position – vector representing mental state of the decision-maker is first laid down on the crime A subspace – evolution of the mental state may result in different position when question regarding guilt of crime B is asked. There have been several examinations of QPT predictions regarding constructive influence and the collapse principle (Pothos & Busemeyer, 2022; White et al., 2014, 2016, 2020; Wojciechowski et al., 2021; Yearsley & Pothos, 2016). It is worth to mention, that following the collapse principle and QPT assumptions it can be predicted, that the more intermediate judgements, the more likely it would be that the judgement would not change but keep collapsing the state back to the initial one, resetting decision-makers view. Yearsley and Pothos (2016) confirmed this effect (called quantum Zeno) experimentally and showed, that frequency of intermediate judgements slows down opinion change.

Comparison of classical and quantum models

Quantum physics was developed to explain puzzling observations that were incomprehensible using classical physical theories. Quantum mechanics required physicists to accept fundamentally new ways of thinking, which ultimately led to a novel understanding of probabilities. Currently, we observe a similar trend in the field of cognitive science. Human judgement frequently deviates from traditional normative standards, and under uncertainty, behaviour can be irrational; there are numerous well-established empirical findings that are difficult to reconcile with classical probability principles (Kahneman, 2012; Tversky, Kahneman, 1974, 1986). Individuals frequently do not provide answers that are consistent with logic or probability theory in response to a variety of cognitive problems (DeNeys, Goel, 2011). In general, the axioms of rational choice are satisfied in transparent circumstances, but they are frequently violated in blurred circumstances (Tversky, Kahneman, 1986). It has been established that individuals do not act in accordance with the assumptions of the expected utility theory (Kahneman, 2012), that they adhere to the sure thing principle, that selecting one option appears to influence subsequent preferences for the available options, producing order effects (White et al., 2016), and that actions do not merely reveal preferences but create them (Ariely & Norton, 2008) Frequently, human judgements are not simply retrieved from memory, but rather are formed based on the cognitive state of the question at hand (Wang et al., 2014).

Prior to the last few years, nearly all cognitive modelling research relied on mathematical principles derived from classical probability theory and classical mechanics. Even though significant progress has been made in understanding cognition in this manner, empirical findings have accumulated that appear puzzling within the framework of classical probability theory. To explain all types of decision fallacies and biases, for instance, various sets of heuristics had to be proposed to bypass or patch the classical theoretical framework in decision research (e.g., Gigerenzer & Todd, 1999; Payne et al., 1993; Tversky & Kahneman, 1974). Quantum theory, in contrast, provides a unified and principled framework for explaining such paradoxical findings and modelling cognition can be reexamined in a new light from quantum theory (e.g., Busemeyer & Bruza, 2012; Busemeyer et al., 2011; Pothos & Busemeyer, 2022).

With the use of QPT conflict, ambiguity and uncertainty can be formalized. Cognitive models based on traditional assumptions assume that at each moment a person is in a definite state with respect to certain emotions, judgement or cognition. However, true state of a person is unknown at each moment to the modeller; therefore, the model can only assign probability to a cognitive response with some value at each moment. The cognitive system is then modelled as if its state follows a well-defined trajectory in its state space. QPT allows a person to be in an indefinite state, experience ambiguity, doubt, conflict at each moment in time. Consequently, it cannot be assumed that psychological states of the subject can be described by definite values, which can be registered in a psychological measurement at randomly chosen moment in time. At each moment, all possible definite values have potential for being expressed. Superposition can be used to reflect uncertainty that people experience in cognitive processes (Wang & Busemeyer, 2013). State of a cognitive system can be modelled as a vector moving (rotating) in its state space over time, until a decision is reached and the state vector collapses to a definite value (Busemeyer et al., 2015; Wang et al., 2013).

Classical cognitive models assume that making measures is a way to capture internal states, feelings, thoughts, or decisions. Measured values are assumed to reflect the state of a cognitive system as it existed immediately before the measurement. Following the traditional approach it is also assumed, that act of measurement itself does not affect observed (recorded) values and if we ask a question – answer will simply reflect the state regarding this question just before we asked it. One of the most surprising lessons learned by physicists from quantum theory is that a measurement of a system both creates and records a property of a system. Niels Bohr emphasized that the outcomes of the quantum measurements must be viewed as the joint product of "system" (state of a subject) and "apparatus" (measuring device). Recorded values are the outcome of interaction of the whole comprehensive experimental setup and not simply a recording of a pre-existing attribute of the measured system. Immediately before asking a question a quantum system is in a superposition. The answer obtained from the system is a consequence of an interaction of the state and the question that has been asked. Definite, sharply defined, specified recorded state of the system is a result of an interaction of the superposition and the measurement. In other words, result of a measurement is a consequence of a constructive interaction between the person's indefinite state and the question being asked. Quantum-like view on measurement is consistent with psychological research, the idea that choice can alter preferences, matches psychological intuition for complex judgements and decisions. Such findings are referred to as "constructive influences" because the conclusion appears to modify (construct) the relevant mental state. There is evidence that even a single judgement can alter the beliefs of its recipients (Ariely & Norton, 2008; Brehm, 1956; Sharot et al., 2010; White et al., 2014; White et al., 2020; Wojciechowski & Pothos, 2018; Wojciechowski et al., 2021).

Classical concepts obey the commutative axiom and as a result it is assumed, that the order of two propositions does not matter when they are considered. Numerous research show, that answering first question can cause a change in the cognitive state and lead a person to different answers to subsequent questions (i.e. Moore, 2002). Sequence of cognitive measurements influences the context of the next measurement and therefore the order of measuring becomes important. This implies that we cannot simultaneously define the joint probability of answers to a conjunction of questions. We can instead only assign a probability to the sequence of responses. In quantum physics, order-dependent measurements are non-commutative operations (Wang & Busemeyer, 2013). Measurement order effects, such as question order effects, are a major concern for attitude, judgement, and decision research, which seeks a quantum-like theoretical understanding of the order effects (e.g., Busemeyer & Wang, 2010; Feldman & Lynch, 1988; Wang & Busemeyer, 2013).

Human cognition and judgement are not always subject to classical rules of logic and probability (Busemeyer & Bruza, 2012; Busemeyer et al., 2011). The classical probabilities employed in contemporary cognitive and decision models are derived from the Kolmogorov axioms (1933), which assign probabilities to sets of events. The Kolmogorov theory's family of sets adheres to the Boolean axioms of logic. The distributive axiom is a fundamental axiom of Boolean logic. This distributive axiom can be used to derive the law of total probability, which forms the basis for Bayes conditional probability. However, many psychological experiments are found to violate the law of total probability (e.g., Tversky & Shafir, 1992). Von Neumann's (1955) axioms, which assign probabilities to events defined as subspaces of a vector space are the basis for QPT. Hughes (1989) found that representing events by subspaces necessitates a logic of subspaces and violates the distributive axiom of Boolean logic (although they form a partial Boolean algebra structure). This demonstrates that quantum models do not always comply with the law of total probability (Busemeyer & Wang, 2015; Busemeyer et al., 2009; Khrennikov, 2010; Pothos & Busemeyer, 2009). Quantum probability is a generalization of probability theory, while quantum logic is a generalization of classical logic. Aerts and Aerts (1995) suggest that the classical probability theory may be insufficient to explain human cognition and judgement in light of the accumulating empirical findings that contradict the classical probabilities principles.

Concepts and processes are assumed to be “decomposable” in cognitive science so that the whole can be understood in terms of its constituent parts. This decomposability is reflected in the common assumption that a complete joint probability distribution exists across all cognitive system measurements, from which one can reconstruct a pairwise joint distribution for any pair of measurements. However, psychological literature provides sufficient evidence to the contrary. For instance, facial cognition is characterized by a strong reliance on the holistic properties of faces (e.g., Farah et al., 1998; Wenger & Townsend, 2001), and it has been argued that perception of ambiguous figures is a holistic process (e.g., Atmanspacher & Filk, 2013; Long & Toppino, 2004). Combinations of concepts cannot be broken down into meaningful components (Aerts et al., 2013; Blutner et al., 2013; Bruza et al., 2009). Quantum models, in contrast to classical models, permit us to describe cognitive systems as non-decomposable, and it may be the case that pairwise probabilities cannot be derived from a common joint probability distribution. Quantum theory refers to these correlations as entanglement correlations (supercorrelations), which are radically different from the classical correlations.

According to Wang and colleagues (2013), there are several fundamental concepts and principles of the QPT resulting in differences between quantum cognitive models and the study of cognitive processes based on the traditional cognitive models. Quantum probability and dispersive states come first. It is essential to understand that the laws governing quantum probabilities are distinct from those governing classical probabilities. Not only are pure quantum superposition and mixed ensemble states dispersive, but also they cannot be represented with points, as is customary in point mechanics. Born’s rule (Born, 1926), which specifies the probability of measuring a particular quantum state based on the square of its amplitude, requires interference factors absent from classical probability theory. It is because of this that quantum probabilities are not additive. Moreover, as states are modified by measurements, the quantum probabilities for transitions between states are heavily dependent on previous measurements. Pure states are the states of quantum systems that are represented as vectors in a vector space (the Hilbert space). Pure states are dispersive, and each pure state superposition consists of another pure state. Mixed states, statistical ensembles of pure states with varying probabilities, constitute a second category of dispersive states. In quantum theory, dispersive pure states are regarded as the origin of “true” (ontic) randomness, whereas the dispersion of mixed states is frequently interpreted as representing (epistemic) partial information about the system’s actual state.

Second, concepts of entanglement, non-local correlations, and the Bell inequalities allow to build cognitive models more accurately, than those based on the assumptions of the classical probability theory. Quantum theory is fundamentally dependent on quantum entanglement. Quantum entanglement refers to a specific kind of quantum superposition. Entangled states exhibit acausal, non-local correlations between the outcomes of measurements performed on geographically distinct subsystems of a system. It suggests that the primary difficulty in comprehending quantum systems is how to decompose them into subsystems, not how to assemble them from basic components. Decomposing a tangled system produces subsystems that are "non-locally" coupled (i.e., without direct causal interactions). Entangled systems can be the result of any interaction. Due to the ubiquitous nature of these correlations in entangled systems, these systems are sometimes referred to as non-separable. Non-local correlations are the basis of quantum holism which are now experimentally well-established in physics, and new applications of them have been originated (Gallus et al., 2023; Vedral, 2006; Waddup et al., 2021).

Third, there can be observables that are both complementary and non-commuting. Quantum systems can have both commuting (such as mass and charge) and non-commuting (such as spin) observables, whereas classical systems are limited to commuting observables only (e.g., position and momentum). As a result of non-commutativity principle, the action sequence B followed by A on an initial state can produce a different result than its reversal. This peculiar property of non-commuting observables in quantum theory is the direct cause of Heisenberg's (1958) uncertainty relations.

Complementarity, non-Boolean logic, and partial Boolean logic are the fourth source of differences. Quantum measurements are formalized in terms of non-classical (non-Boolean) propositional logic (Birkhoff & von Neumann, 1936). In contrast to the two-valued (yes/no) logic of classical binary alternatives, in which a proposition's complement and negation are identical, the complement of a quantum proposition is frequently distinct from its negation. In the Boolean sense, disbelief is both the complement and negation of belief, whereas in the quantum non-Boolean sense, doubt and inquiry are complements of belief but not negations.

Quantum logic is inconsistent with the Boolean law of the excluded middle, also known as the *tertium non-datur*. According to an early version of quantum logic (Birkhoff & von Neumann, 1936), this is a violation of the complement law of distributivity.

Non-commutativity of quantum observables corresponds immediately to non-distributivity of the quantum proposition lattice. Both formal criteria demonstrate that non-commuting observables and their accompanying statements cannot coexist. However, it is still possible to discuss the compatibility or incompatibility of concepts despite the difficulty of analysing such minor degrees of incompatibility in the context of the definition of complementarity given above. The propositional logic framework is more general than the algebraic framework because it can be applied to situations where quantum theory operators are undefined. Partial Boolean logic is an appropriate logical framework for systems that exhibit both classical and quantum behaviour. Partial Boolean logic consists of "locally" Boolean sublattices that are glued together in a globally non-Boolean fashion. These structures are associated with algebras of observables containing both non-commuting and commuting observables.

Boundary between classical and quantum probability theory

According to Penrose (2016), there are fundamental difference differences between the classical and quantum physics. Classical view is based on the assumption that there exists an objective world, which evolves in a clear and deterministic way. The world is being governed precisely by formulated mathematical equations, governing its evolution. Classical physics is in accordance with common sense, assumes that physical reality exists independently of the observer, and is not affected by how we might choose to look at it. Physical experiments showed however that this world view is not true.

Quantum theory was not wished upon by physicists, and it the view of the quantum world following laboratory findings was met with great reluctance. According to the quantum theory, the state cannot be described as it "actually is" but can be only defined as a result of certain measurements. Quantum mechanics is often described as the physics of very small (atoms, molecules or subatomic particles). For many decades, concepts such as entanglement and superposition have been associated with quantum mechanics only and different language was used to describe quantum and macroscopic level. However, there is a growing number of experiments showing, that non-quantum entanglement can exist and there are examples of superposition on a macroscopic scale (Gallus et al., 2023; Qian et al., 2015; Shalyt-Margolin, 2021).

Physicists are not only discussing assumptions and different interpretations of quantum theory, but still exploring borders between the classical and quantum theory (Aerts et al., 2019; Khrennikov, 2010, 2014; Milovanović & Saulig, 2022; Penrose, 2004; Slavnov, 2006). It is generally acknowledged that classical physics is deterministic. Restating a famous argument attributed to Laplace (known as Laplace's demon), determinism is commonly understood to be the view that sufficient knowledge of the laws of nature and appropriate boundary conditions will allow a superior intelligence to predict the future states of the physical world and to retrodict its past states with infinite precision. Quantum theory is considered as the first theory to have introduced genuine indeterminism into physics. Doubts about classical determinism were fostered in the second half of the 20th century, after development of the theory of chaotic systems and understanding of its implications. Del Santo and Gisin (2019; see also Gisin, 2021) challenged the view that there is a fundamental difference between quantum and classical physics, built on the notion of determinism. Arguing that physical quantities do not have an actual value and that it cannot be measured with infinite precision, they undermine the principle of infinite precision and present possible, quantum-based interpretation of classical physics. The novel model proposed by the researchers rejects the conventional physical interpretation of mathematical real numbers (numbers with an unlimited number of specified digits). Instead, it claims that after a given number of digits, their values become really random and that only their tendency to take a certain value is well-defined. This results in new understandings on the link between classical and quantum physics. When, how and under what conditions an indeterminate quantity becomes a definite value is a well-known topic in the foundations of quantum physics known as the quantum measurement problem. This is because, in the quantum universe, it is impossible to see reality without altering it. In actuality, the value of a measurement performed on a quantum item is not determined until an observer performs the measurement. According to Del Santo and Gisin (2019), the same problem may have always been concealed by the soothing laws of classical physics. In Penrose's opinion (2016) quantum and macroscopic worlds should be treated as one, we cannot draw a line separating one from the other. Classical physics should be treated as a good approximation when a large number of quantum particles are participating in a process and quantum theory is best to describe mechanisms of elementary particle's behaviour.

Some of the founding fathers of quantum theory, most ardently Niels Bohr, were convinced that the theory's central notions would prove meaningful outside of physics, in areas such as psychology and philosophy. Several authors have conducted studies aiming to determine whether classical or QPT allows for a more accurate description of results of cognitive experiments. Busemeyer and Wang (2018) conducted two large-scale investigations into measurement context effects. In one study, participants were provided with public health service announcements and asked to choose between two options for six pairs of questions rating the announcements' effectiveness. In a separate study, participants viewed computer-generated avatars and answered six pairs of questions based on four evaluations of their attitudes toward the avatars using binary choices. In all investigations, Busemeyer and Wang (2018) demonstrated that quantum probability models performed better than Bayesian causal network models in predicting the joint frequencies of the six two-way table measures. White and colleagues (2014, 2016; Wojciechowski et al., 2022) in a series of experiments also confirmed, that cognitive models based on QPT are better suited to empirical data.

The quantum probability models for conceptual combinations, entanglement and super correlations have also been used (Aerts et al., 2017). Probabilistic dynamic systems in psychology are typically characterized with Markov models, such as random walk/diffusion models of decision-making (Ratcliff et al., 2016). Quantum theory offers an additional generic paradigm for probabilistic dynamic systems. In such tasks, the classical Markov and quantum models are comparable, but there is a crucial distinction when measurements are done sequentially through time. The Chapman–Kolmogorov equation is a form of law of total probability throughout time that the Markov model obeys, whereas the quantum probability model permits violations of this law. Kvam et al. (2015) evaluated the two types of dynamic models using a perceptual judgement paradigm in which participants were required to determine the main motion direction in a dynamic dot display. In a choice–confidence condition, participants examined the dots, made a decision regarding motion, reexamined the dots, and then rated their level of confidence. In a confidence-only condition, participants just observed the dots and then rated their level of confidence. According to the Chapman–Kolmogorov equation, the distribution of confidence ratings across the two situations should have been the same, but this was not the case. According to the quantum model, the difference between the conditions suggested an interference effect. In addition, Kvam et al.'s (2015) quantum model produced better fits than a corresponding Markov model.

A second crucial distinction between Markov and QPT models is the manner in which dynamics evolve over time. Conventional Markov models of evidence accumulation predict a monotonic growth in the mean of the accumulation over time, which makes numerous preference reversals possible only in multiattribute choice and, even then, reliant on the specific manner the available options are compared. In contrast, QPT dynamics typically predict systematic oscillation effects over time. Hence, quantum dynamical models are applicable when we anticipate ambivalence and reflection during the decision-making process. Broekaert and colleagues (2020) showed that model based on quantum dynamics can describe order effect in relation to disjunction effect and inflation effect on gambling more fully and accurately than the Markov model.

Kvam and Busemeyer (2020) used a preference task to test these hypotheses. Participants were given a choice between two coupons for restaurants that varied in quality, price and proximity. Participants were requested preference ratings at multiple time points, both with and without a prior selection. According to Kvam et al. (2015), the choice made beforehand created interference effects. In addition, the scientists discovered that preferences oscillated predictably over time, as expected by a QPT model. Kvam et al. (2015) contrasted the predictions for choice and confidence from a quantum walk model with a Markov random walk model using a Bayes factor, and for the majority of participants, the Bayes factor favoured the quantum model. Busemeyer and Wang (2019) utilized a generalization test to compare the two models: two time intervals for assessing confidence were used to estimate the model parameters, which were then utilized to forecast the confidence ratings for a third time interval. The quantum walk model generated more accurate predictions for generalizations. Using choice and reaction time distributions, Markov and quantum walk models have also been contrasted. Fuss and Navarro (2013) discovered that a quantum dynamic model has a little advantage over a Markov random walk model for forecasting choice and response time in a signal detection challenge. The cyclical nature of QPT dynamics provides a precise embodiment of the notion in social psychology that humans can go through multiple cycles of back-and-forth before reaching a conclusion. Results of study conducted by Trueblood and colleagues (2017) indicate that QPT thinking emerges when participants are less familiar with a problem, and respond more reflexively. QPT may be considered as the default mode of thought, gradually replaced by classical probability theory, depending on the initial significance of the task and as experience with a task accumulates.

The aforementioned results of the experiments indicate that QPT developed for physics of the elementary particles are also suitable for macroscopic objects and analysis of human cognition. In the later chapters of this book, you can find numerous examples of studies confirming the superiority of QPT over classical probability theory in cognition research. In conclusion to this section it can be said, following Pothos and Busemeyer (2022), that in the context of a single random variable, the quantum approach and classical method will provide similar results. However, it can be expected that in the presence of several random variables, quantum and traditional methods can provide unique outcomes. This difference permits quantum models to account for the contextual influences on judgement. This context sensitivity enables quantum decision-making to account for and anticipate heuristics and biases that are challenging to reconcile with rational choice models (Jaeger & Trueblood, 2019; Pothos & Busemeyer, 2022; Wang et al., 2013).

Current quantum cognitive models

Recently, models of decision-making based on the mathematics of quantum theory have garnered a great deal of interest (Aerts, 2009; Busemeyer & Bruza, 2012; Khrennikov, 2010, 2014; Mogiliansky et al., 2009; Pothos & Busemeyer, 2013, 2022; Wang et al., 2013; Yukalov & Sornette, 2010, 2014). In part, these models represent a response to the empirical issues faced by "rational" decision-making models, such as those based on Bayesian probability theory (such instances are typically connected with the well-known Tversky–Kahneman research tradition) (Chater et al., 2008; Tversky & Kahneman, 1974). These quantum models propose that, at least under some conditions, human behaviour is inconsistent with classical probability theory and predicted utility maximization. In contrast to, for instance, the quick and frugal heuristics program (Gigerenzer & Goldstein, 1996), quantum cognition intends to replace the framework of conventional probability theory with an alternative theory of probabilities. This new probability theory contains characteristics, such as context effects, interference effects, and constructive judgements, that are consistent with psychological intuition on human decision-making.

Initial research involving quantum models tended to concentrate on explaining paradoxical results from the perspective of classical probability theory, and there have been a number of successes in this area (Aerts et al., 2013; Blutner at al., 2013; Bruza et al., 2015; Pothos & Busemeyer, 2009, 2013; Trueblood & Busemeyer, 2011; Wang et al., 2013; White et al., 2014). Aerts and colleagues (e.g., Aerts & Aerts, 1995; Aerts et al., 2019) took the initial steps to implement Bohr's suggestion in the 1990s. To examine quantum-like behaviour in non-quantum systems, they employed non-distributive propositional lattices. Khrennikov (1997), focused on non-classical probabilities, and Atmanspacher et al. (2002) on utilizing an algebraic framework based on non-commuting operations with fruitful results, introduced alternative methods. Since 2006, Busemeyer and colleagues have used QPT to successfully address concrete and long-standing paradoxical problems in decision and cognition research (Busemeyer & Bruza, 2012; Busemeyer & Wang, 2010; Busemeyer et al., 2011).

The rapid development of the quantum cognition research program has centred on the few areas, starting with the decision-making processes. A forerunner is attributable to Aerts and Aerts (1995). Busemeyer, Wang et al. (2009), Pothos and Busemeyer (2009, 2013), Yukalov and Sornette (2010), and Navarro and Fuss (2009) developed detailed applications. The central concept is to define decision-making process in terms of quantum probability amplitudes in a Hilbert space model. Long-standing challenges in the literature on decision-making, such as the disjunction effect (Tversky & Shafir, 1992), have been explained by quantum interference (Busemeyer et al., 2006; Khrennikov & Haven, 2009; Pothos & Busemeyer, 2009; Yukalov & Sornette, 2010). Busemeyer and Bruza (2012) and others have applied quantum logic in the form of closed subspaces of a Hilbert space to a variety of cognitive phenomena.

The bistable perception of ambiguous figures, such as the Necker cube, is a good illustration of the uncertain perception. Atmanspacher et al. (2004) and Atmanspacher and Filk (2010) created a thorough model defining several experimentally confirmed psychophysical characteristics of bistable perception. In addition, Atmanspacher and Filk (2010, 2013) anticipated that distinct states in bistable perception may violate the temporal Bell inequalities, a litmus test for quantum behaviour. Conte and colleagues (2009) have also conducted studies applying quantum theory to the perception of ambiguous figures.

The complex subject of meaning in natural languages is frequently investigated using semantic networks. Gabora and Aerts (2002) outlined the contextual manner in which concepts are invoked, employed and combined to produce meaning. Their theories on concept connection in an evolutionary framework have been refined in recent research (Aerts et al., 2013; Gabora & Aerts, 2009). Bruza et al. (2009) investigated meaning relationships in terms of entanglement-style characteristics in quantum representations of the human mental lexicon and presented related experimental testing. Relevant work is owed to Blutner and coworkers (Blutner et al., 2013). And Gabora and Kitto (2017) proposed that cognitive humour can be modelled using the mathematical framework of quantum theory and presented the Quantum Theory of Humor.

Recently, there has been a shift in emphasis towards verifying novel predictions derived from quantum models and devising improved tests of quantum versus classical decision theories (Atmanspacher & Filk, 2010; Yearsley & Pothos, 2022). The treatment of question order effects is a major achievement of quantum models of cognition (Moore, 2002). The term "order effects" refers to a phenomena in which the predicted distribution of replies to a single question varies on whether it was asked first or second in a series of questions. In other words, the outcome of a subsequent question can be affected by an earlier question. According to newly developed model, order effects are a natural consequence of quantum theory and may thus be explained by quantum cognitive models. Quantum theory also predicts special limits on the probabilities that can be created by these models, most notably in the form of the quantum question (QQ) Equality, which is even more astounding (Wang & Busemeyer, 2013; Wang et al., 2014). Data from real-world trials and surveys collected by Wang and colleagues indicate that these limitations are extremely well met (Wang et al., 2014). QQ equality provides a simple and very precise account for a surprising regularity regarding measurement order effects in social and behavioural sciences. Question order effects are both a natural application of quantum theory and a remarkable empirical support of the concept of utilizing quantum theory to explain decisions. Pothos and colleagues (2013) proposed the Quantum Similarity Model (QSM) to represent conjunction fallacy and the asymmetries in similarity judgements. Epping and colleagues (2023) extended the QSM to a more general framework that can be fitted to similarity judgements.

Human irrationality has been blamed for a number of strange things that people do that can't be explained by classical probability. The conjunction fallacy occurs when individuals incorrectly conclude that the likelihood of events A plus B is greater than the chance of event A alone (Tversky & Kahneman, 1983). The disjunction fallacy occurs when people assess the likelihood of A or B to be less than the probability of A alone (Carlson & Yates, 1989). Busemeyer et al. (2011) created a single quantum model to account for these "errors" in probability judgement. Moreover, Trueblood and Busemeyer (2011) proposed a quantum inference model to explain order effects that may occur in sequential probability assessment tasks. Many analyses have been conducted on QPT predictions involving constructive impacts and the collapse principle (Pothos & Busemeyer, 2022). Moreira and Wichert (2016) explored a quantum structure to perform quantum probabilistic interferences to accommodate the paradoxical findings of the sure thing principle and proposed a quantum-like Bayesian network which allows to replace classical probabilities with quantum probability amplitudes. White et al. (2014, 2020) utilized pairs of stimuli with significant positive or negative valence, in which the second stimulus was always rated, but the first stimulus was rated only 50% of the time. They expected, based on a simple QPT model, that rating the first stimulus would heighten the evaluation of the second stimulus (e.g., make it more positive if the stimulus was already positive). This prediction is referred to as an evaluation bias, and it has been frequently validated (see also: Wojciechowski et al., 2022). Busemeyer and Wang (2018) developed the Hilbert space multidimensional theory (HSM theory), a general theory of measurement context effects in the social and behavioural sciences. HSM theory provides a way to model context effects by using QPT, with all the variables represented within low dimensionality vector space.

Long ago acknowledged, but still inadequately understood today, are question order effects in surveys (Schwarz & Sudman, 1992). To empirically evaluate the quantum question order model, Wang and Busemeyer (2013) extracted an *a priori*, accurate, quantitative prediction about the order effects of attitude questions from quantum theory, which is utilized to rigorously test the hypothesis. In addition, Atmanspacher and Römer (2012) projected unique types of order effects and pointed out possible limitations of Hilbert space representations.

Yearsley and Pothos (2016) investigated a prediction derived from the collapse principle known as the quantum Zeno effect. In their experiment, participants were asked to determine the chance that a defendant was guilty of a crime based on a series of facts; the defendant was initially assumed to be innocent. In different conditions, participants were asked to make assessments after receiving each of numerous pieces of information, varying between conditions. According to the collapse principle, the more intermediate judgements were expressed, the more likely it was that the judgements would keep collapsing the state back to the initial one, so resetting the person's belief. This is the quantum Zeno effect, which in a decision-making context refers to the notion that the frequency of intermediate judgements slows down the rate of opinion change (colloquially, a watched pot never boils). Quantum models with bistable perception tasks have also been used to examine Zeno effects (Atmanspacher et al., 2010). Another intriguing application of the collapse principle is the so-called zero priors dilemma. A stipulation of Bayes' rule in classical probability theory is that if a person assigns a probability of zero or one to any hypothesis, then no additional evidence may modify the likelihood from these extremes. Studies show that this rule conflicts with human judgements. Basieva et al. (2017) conducted two tests in which participants assessed the likelihood of guilt for a variety of potential criminals. Initial evidence led participants to assign zero (or extremely low) probability to one candidate, but subsequent evidence shifted these zero prior probabilities to a high degree of guilt. In Basieva et al.'s (2017) quantum model, it was show potential of a change from zero priors to a nonzero probability of guilt by focusing on the way a previously made decision alters the mental state.

Observations that are confusing from a traditional probability standpoint are also present in more fundamental cognitive processes, such as memory processing. The episodic overdistribution effect is a paradoxical occurrence that can be explained by a simple quantum model involving the superposition of memory traces (Brainerd et al., 2013). Bruza and colleagues (2009) created a quantum model to explain the entanglement-like behaviour observed in the human mental lexicon and word association. Trueblood and Hemmer (2017) proposed a generalized Quantum Episodic Memory Model, explaining episodic overdistribution effect and subadditivity. Yearsley and Pothos (2014) explored temporal Bell (Leggett–Garg) inequalities and presented an alternative conceptualization of time in cognition. In one of the most recent papers, Waddup and colleagues (2023) investigated the memory and specifically retrieval processes using temporal Bell inequality to test sensitivity of sequential measurement contexts. The findings support recent studies on quantum-like representations in memory and challenge their adaptive value.

These ongoing research fields demonstrate the promise for formalizing a vast array of cognitive events using the coherent ideas of quantum theory. From a conventional probability perspective, these behaviours can appear strange. Quantum theory is innately congruent with deeply embedded psychological ideas and intuitions, perhaps contradicting the prevalent perception that it is strange. However, it fits the data collected in numerous studies and provides a novel conceptual framework for elucidating empirical difficulties of cognition and a wealth of new formal tools for cognitive modelling. In the following chapters, we will apply some of the models developed to date to legal decision-making.

Although the mathematical machinery of quantum theory is not inherently more sophisticated than that required by many other cognitive models, consisting primarily of linear algebra and a tiny amount of calculus, the majority of cognitive scientists are unfamiliar with it. Busemeyer and Bruza (2012) wrote an introduction of the application of quantum theory to cognitive and decision scientists. Yearsley and Busemeyer (2016) prepared a tutorial to give an introduction to quantum models, with a particular emphasis on how to build quantum-like models in practice. General procedures for constructing, estimating and testing models based on the Hilbert space can be also found in Busemeyer's and Wang's works, chapter by Busemeyer, Wang and Pothos (2015), books by Khrennikov (1997, 2010, 2014), Haven and Khrennikov (2013), Narens (2007), Batchelder et al. (2017) and Aerts et al. (2019). Books by Penrose (2004, 2016), Baggott (2020), Dirac (1964), Wheeler and Zurek (1983) and Shankar (1994) are worthy introductions to quantum theory.

References

Aerts, D. (2009). Quantum structure in cognition. *Journal of Mathematical Psychology*, Vol. 53, No. 5, pp. 314–348; DOI:10.1016/j.jmp.2009.04.005.

Aerts, D., Aerts, A., Beltran, L., Geriente, S., Sassoli de Bianchi, M., Sozzo, S., & Veloz, T. (2017). Spin and wind Directions I: Identifying entanglement in nature and cognition. *Foundations of Science*, Vol. 23, pp. 323–335; DOI: 10.1007/s10699-017-9528-9.

Aerts, D., & Aerts, S. (1995). Applications of quantum statistics in psychological studies of decision processes. *Foundations of Science,* Vol. 1, pp. 85–97.

Aerts, D., Gabora, L., & Sozzo, S. (2013). Concepts and their dynamics: A quantum – Theoretic modeling of human thought. *Topics in Cognitive Science*, Vol. 5, No. 4, pp. 737–772; DOI: 10.1111/tops.12042.

Aerts, D., Khrennikov, A., Melucci, M., & Toni, B. (2019). *Quantum – Like Models for Information Retrieval and Decision – Making*. Springer, Cham.

Ariely, D., & Norton, M.I. (2008). How actions create – Not just reveal – Preferences. *Trends in Cognitive Sciences*, Vol. 12, pp. 13–16; DOI: 10.1016/j.tics.2007.10.008.

Atmanspacher, H. & Filk, T. (2010). A proposed test of temporal nonlocality in bistable perception. *Journal of Mathematical Psychology*, Vol. 54, No. 3, pp. 314–321; DOI: 10.1016/j.jmp.2009.12.001.

Atmanspacher, H., Filk, T., & Roemer, H. (2004). Quantum Zeno features of bistable perception. *Biological Cybernetics*, Vol. 90, pp. 33–40; DOI: 10.1007/s00422-003-0436-4.

Atmanspacher, H., Römer, H., & Walach, H. (2002). Weak quantum theory: Complementarity and entanglement in physics and beyond. *Foundations of Physics*, Vol. 32, No. 3, pp. 379–406; DOI: 10.1023/A:1014809312397.

Atmanspacher, H., & Römer, H. (2012). Order effects in sequential measurements of non-commuting psychological observables. *Journal of Mathematical Psychology*, Vol. 56, No. 4, pp. 274–280; DOI: 10.1016/j.jmp.2012.06.003.

Baggott, J. (2020). *The Quantum Cookbook. Mathematical Recipes for the Foundations of Quantum Mechanics.* Oxford University Press: Oxford.

Basieva, I., Pothos, E.M., Trueblood, J., Khrennikov, A., & Busemeyer, J.R. (2017). Quantum probability updating from zero prior (by-passing Cromwell's rule). *Journal of Mathematical Psychology*, Vol. 77, pp. 58–69; DOI: 10.1016/j.jmp.2016.08.005.

Batchelder, W.H., Colonius, H., Dzhafarov, E.N., & Myung, J. [eds.] (2017). *New Handbook of Mathematical Psychology*. Cambridge University Press, Cambridge.

Birkhoff, G., & von Neumann, J. (1936). The logic of quantum mechanics. *Annals of Mathematics*, Vol. 37, pp. 823–843.

Blutner, R., Pothos, E.M., & Bruza, P. (2013). A quantum probability perspective on borderline vagueness. *Topics in Cognitive Sciences*, Vol. 5, pp. 711–736; DOI: 10.1111/tops.12041.

Born, M. (1926). On Quantum Mechanics II. *Physics*, Vol. 35, pp. 557–615.

Brainerd Ch.J., Wang, Z., & Reyna, V.F. (2013). Superposition of episodic memories: Overdistribution and quantum models. *Topics in Cognitive Sciences*, Vol. 5, pp. 773–799; DOI: 10.1111/tops.12039.

Brehm, J.W. (1956). Post-decision changes in the desirability of choice alternatives. *Journal of Abnormal and Social Psychology*, Vol. 52, pp. 384–389; DOI: 10.1037/h0041006.

Broekaert, J.B., Busemeyer, J.R., & Pothos, E.M. (2020). The disjunction effect in two-stage simulated gambles. An experimental study and comparison of a heuristic logistic, Markov and quantum-like model. *Cognitive Psychology*, Vol. 117; DOI: 10.1016/j.cogpsych.2019.101262.

Bruza, P.D., Kitto, K., Nelson, D., & McEvoy, C. L. (2009). Is there something quantum-like about the human mental lexicon? *Journal of Mathematical Psychology*, Vol. 53, pp. 362–377; DOI: 10.1016/j.jmp.2009.04.004.

Bruza, P.D., Wang, Z.J., & Busemeyer, J.R. (2015). Quantum cognition: A new theoretical approach to psychology. *Trends in Cognitive Sciences*, Vol. 19, No. 7, pp. 383–393; DOI: 10.1016/j.tics.2015.05.001.

Busemeyer, J.R., & Bruza, P.D. (2012). *Quantum Models of Cognition and Decision*. Cambridge University Press, Cambridge.

Busemeyer, J.R., Pothos, E.M., Franco, R., & Trueblood, J.S. (2011). A quantum theoretical explanation for probability judgment errors. *Psychological Review*, Vol. 118, No. 2, pp. 193–218; DOI: 10.1037/a0022542.

Busemeyer, J.R., Wang, Z., & Lambert – Mogiliansky, A. (2009). Empirical comparison of Markov and quantum models of decision making. *Journal of Mathematical Psychology*, Vol. 53, No. 5, pp. 423–433; DOI: 10.1016/j.jmp.2009.003.002.

Busemeyer, J.R., & Wang, Z. (2015). What is quantum cognition, and how is it applied to psychology? *Current Directions in Psychological Science*, Vol. 24, No. 3, pp. 163–169; DOI: 10.1177/0963721414568663.

Busemeyer, J.R., Wang, Z., & Pothos, E. (2015). Quantum models of cognition and decision, [in:] *Oxford Handbook of Computational and Mathematical Psychology*, [eds.] Busemeyer, J.R., Wang, Z., Townsend, J.T., & Eidels, A., Oxford University Press, New York, pp. 369–389.

Busemeyer, J.R., Wang, Z., Pothos, E., & Trueblood, J.S. (2015). The conjunction fallacy, confirmation and quantum theory: Comment on Tentori, Crupi and Russo (2013). *Journal of Experimental Psychology: General*, Vol. 144, No. 1, pp. 236–243; DOI: 10.1037/xge0000035.

Busemeyer, J.R., Wang, Z., & Townsend, J.T. (2006). Quantum dynamics of human decision-making. *Journal of Mathematical Psychology*, Vol. 50, pp. 220–241; DOI: 10.1016/j.jmp.2006.01.003

Busemeyer, J.R., Wang, Z., Townsend, J.T., & Eidels, A. (2015). *The Oxford Handbook of Computational and Mathematical Psychology*. Oxford University Press, Oxford.

Busemeyer, J.R., & Wang, Z. (2018a). Hilbert space multidimensional theory. *Psychological Review*, Vol. 125, No. 4, pp. 572–591; DOI: 10.1037/rev0000106.

Busemeyer, J.R., & Wang, Z. (2018b). Data fusion using Hilbert space multi – dimensional models. *Theoretical Computer Science*, Vol. 752, pp. 41–55; DOI: 10.1016/j.tcs.2017.12.007

Busemeyer, J.R., & Wang, Z. (2019). Introduction to Hilbert space multi – dimensional modeling, [in:] *Quantum-Like Models for Information Retrieval and Decision – Making, STEAM – H: Science, Technology, Engineering, Agriculture, Mathematics & Health*, [eds.] Aerts, D., et al., Springer, Cham, pp. 41–49. DOI: 10.1007/978-3-030-25913-6_3

Carlson, B.W., & Yates, J.F. (1989). Disjunction errors in qualitative likelihood judgment. *Organizational Behavior and Human Decision Processess*, Vol. 43, No. 3, pp. 368–379; DOI: 10.1016/0749–5978(89)90014-9.

Chater, N., Vlaev, I., & Grinberg, M. (2008). A new consequence of Simpson's paradox: Stable cooperation in one-shot prisoner's dilemma from populations of individualistic learners. *Journal of Experimental Psychology: General*, Vol. 137, pp. 403–421; DOI: 10.1037/0096–3445.137.3.403.

de Finetti, B., Machi, A., & Smith, A. (1993). *Theory of Probability: A Critical Introductory Treatment*. Wiley, New York.

Del Santo, F., & Gisin, N. (2019). Physics without determinism: Alternative interpretations of classical physics. *Physical Review A*, Vol. 100, No. 062107; DOI: 10.1103/PhysRevA.100.062107.

DeNeys, W., & Goel, V. (2011). Heuristics and biases in the brain: Dual neural pathways for decision making, [in:] *Neuroscience of Decision Making*, [eds.] Vartanian, O., Mandel, D.R., Psychology Press, New York, pp. 125–141.

Dirac, P. (1964). *Lectures on Quantum Mechanics*. Dover Publications, New York.

Dulany, D.E., & Hilton, D. (1991). Conversational implicature, conscious representation, and the conjunction fallacy. *Social Cognition*, Vol. 9, pp. 85–110; DOI: 10.1521/soco.1991.9.1.85.

Elqayam, S., & Evans, J.S.B.T. (2013). Rationality in the new paradigm: Strict versus soft Bayesian approaches. *Thinking and Reasoning*, Vol. 19, pp. 453–470; DOI: 10.1080/13546783.2013.834268.

Epping, G.P., Fisher, E.L., Zeleznikow-Johnston, A.M., Pothos, E.M., & Tsuchiya, N. (2023). A quantum geometric framework for modeling color similarity judgments. *Cognitive Science*, Vol. 47, No. 1, p. e13231; DOI: 10.1111/cogs.13231.

Farah, M.J., Wilson, K.D., Drain, M., & Tanaka, J.N. (1998). What is "special about face perception? *Psychological Review*, Vol. 105, No. 3, pp. 482–498; DOI: 10.1037/0033–295X.105.3.482.

Feldman, J.M., & Lynch, J.G. (1988). Self – generated validity and other effects of measurement on belief, attitude, intention, and behavior. *Journal of Applied Psychology*, Vol. 73, No. 3, pp. 421–435; DOI: 10.1037/0021–9010.73.3.421.

Fuss, I.G. & Navarro, D.J. (2013). Open parallel cooperative and competitive decision processes: A potential provenance for quantum probability decision models. *Topics in Cognitive Science*, Vol. 5, No. 4, pp. 818–843; DOI: 10.1111/tops.12045

Gabora, L., & Aerts, D. (2002). Contextualizing concepts using a mathematical generalization of the quantum formalism. *Journal of Experimental and Theoretical Artificial Intelligence*, Vol. 14, No. 4, pp. 327–358; DOI: 10.1080/09528130210162253.

Gabora, L., & Kitto, K. (2017). Toward a quantum theory of humor. *Frontiers in Psychology*, Vol. 4; DOI: 10.3389/fphy.2016.00053.

Gallus, C., Pothos, E.M., Blasiak, P., Yearsley, J.M., & Wojciechowski, B.W. (2023). Bell correlations outside physics. *Scientific Reports*, Vol. 13, No. 4394; DOI: 10.1038/s41598-023-31441-x.

Gigerenzer, G., Goldstein, D. (1996). Reasoning the fast and frugal way: Models of bounded rationality. *Psychological Review*, Vol. 103, pp. 650–669; DOI: 10.1037/0033-295x.103.4.650.

Gigerenzer, G., & Todd, P.M. (1999). Fast and frugal heuristics: The adaptive toolbox, [in:] *The ABC Research Group, Simple Heuristics That Make us Smart*, [eds.] Gigerenzer, G., & Todd, P.M., Oxford University Press: Oxford, pp. 3–34.

Gisin, N. (2021). Indeterminism in physics, classical chaos and bohmian mechanics: Are real numbers really real? *Erkenntnis*, Vol. 86, pp. 1469–1481; DOI: 10.1007/s10670-019-00165-8.

Haven, E., & Khrennikov, A. (2013). *Quantum Social Science*. Cambridge University Press, Cambridge.

Heisenberg, W. (1958). *Physics and Philosophy: The Revolution in Modern Science, Role of Modern Physics in Human Thinking*. Harper & Row, New York.

Howson, C., & Urbach, P. (1993). *Scientific Reasoning: The Bayesian Approach*. Open Court, Chicago.

Hughes, R.I.G. (1989). *The Structure and Interpretation of Quantum Mechanics*. Harvard University Press, Cambridge, MA.

Jaeger, Ch.B., & Trueblood, J.S. (2019). Thinking quantum: A new perspective on decisionmaking in law. *Florida State University Law Review*, Vol. 46, No. 4, pp. 733–806.

Kahneman, D. (2001). *Thinking Fast and Slow*. Penguin, London.

Kahneman, D. (2012). *Pułapki Myślenia. O Myśleniu Szybkim I Wolnym*. Media Rodzina, Poznań.

Khrennikov, A. (1997). *Non – Archmiedean Analysis: Quantum Paradoxes, Dynamical Systems and Biological Models*. Kluwer Academic Publishers, Dordrecht.

Khrennikov, A. (2010). *Ubiquitious Quantum Structure. From Psychology to Finance*. Springer, Heidelberg.

Khrennikov, A. (2014). *Beyond Quantum*. CRC Press, Boca Raton.

Kolmogorov, A.N. (1933). *Foundations of Probability Theory*. Julius Springer, Berlin.

Kvam, P.D., Busemeyer, J.R. (2020). A distributional and dynamic theory of pricing and preference. *Psychological Review*, Vol. 127, No. 6, pp. 1053–1078; DOI: 10.1037/rev0000215.

Kvam, P.D., Pleskac, T.J., Yu, S., & Busemeyer, J.R. (2015). Interference effects of choice on confidence: Quantum characteristics of evidence accumulation. *PNAS*, Vol. 112, No. 34, pp.10645–10650; DOI: 10.1073/pnas.1500688112.

Long, G.M., & Toppino, T.C. (2004). Enduring interest in perceptual ambiguity: Alternating views of reversible figures. *Psychological Bulletin*, Vol. 130, No. 5, pp. 748–768; DOI: 10.1037/0033–2909.130.5.748.

Łukasik, A. (2017). *Filozoficzne Zagadnienia Mechaniki Kwantowej*. Wydawnictwo UMCS, Lublin.

Milovanovic, M., & Saulig, N. (2022). An intensional probability theory: Investigating the link between classical and quantum probabilities. *Mathematics*, Vol. 10, No. 22, 4294; DOI: 10.3390/math10224294.

Moore, D.W. (2002). Measuring new types of question-order effects. *Public Opinion Quarterly*, Vol. 66, No. 1, pp. 80–91; DOI: 10.1086/338631.

Moreira, C., & Wichert, A. (2016). Quantum – like Bayesian networks for modelling decision making. *Frontiers in Psychology*, Vol. 7, No. 11; DOI: 10.3389/fpsyg.2016.00011.

Moro, R. (2009). On the nature of the conjunction fallacy. *Synthese*, Vol. 171, pp. 1–24; DOI: 10.1007/s11229.008.9377.8.

Narens, L. (2007). *Introduction to the Theories of Measurement and Meaningfulness and the Use of Symmetry in Science*. Lawrence Erlbaum Associates Publishers, Mahwah.

Navarro, D.J., & Fuss, I.G. (2009). Fast and accurate calculations for first – passage times in Wiener diffusion models. *Journal of Mathematical Psychology*, Vol. 53, No. 4, pp. 222–230; DOI: 10.1016/j.jmp.2009.02.003.

Oaksford, M., & Chater, N. (2009). Précis of Bayesian rationality: The probabilistic approach to human reasoning. *Behavioral and Brain Sciences*, Vol. 32, pp. 69–120; DOI: 10.1017/S0140525X09000284.

Payne, J.W., Bettman, J.R., & Johnson, J. (1993). *The Adaptive Decision Maker*. Cambridge University Press, Cambridge.

Penrose, R. (2004). *The Road to Reality. A Complete Guide to the Laws of the Universe.* Jonathan Cape, London.

Penrose, R. (2016). *The Emperor's New Mind: Concerning Computers, Minds, and the Laws of Physics.* Oxford University Press, Oxford.

Pothos, E.M., & Busemeyer, J.R. (2009). A quantum probability explanation for violations of "rational" decision theory. *Proceedings of the Royal Society B: Biological Sciences*, Vol. 276, No. 1665, pp. 2171–2178; DOI: 10.1098/rspb.2009.0121.

Pothos, E.M., & Busemeyer, J.R. (2013). Can quantum probability provide a new direction for cognitive modeling? *Behavioral and Brain Sciences*, Vol. 36, pp. 255–327; DOI: 10.1017/S0140525X12001525.

Pothos, E.M., Busemeyer, J.R., & Trueblood, J.S. (2013). A quantum geometric model of similarity. *Psychological Review*, Vol. 120, No. 3, pp. 679–696; DOI: 10.1037/a0033142.

Pothos, E.M., Busemeyer, J.R., Shiffrin, R.M., & Yearsley, J.M. (2017). The rational status of quantum cognition. *Journal of Experimental Psychology: General*, Vol. 146, No. 7, pp. 968–987; DOI: 10.1037/xge0000312.

Pothos, E.M., & Busemeyer, J.R. (2022). Quantum cognition. *Annual Reviews in Psychology*, Vol. 73, pp. 749–778; DOI: 10.1146/annurev-psych-033020–123501.

Qian, X.F., Little, B., Howell, J.C., & Eberly, J.H. (2015). Shifting the quantum – classical boundary: Theory and experiment for statistically classical optical fields: Supplementary material. *Optica*, Vol. 2, No. 7, pp. 611–615; DOI: 10.1364/OPTICA.2.000611.

Ratcliff, R., Smith, P.L., Brown, S.L., & McCoon, G. (2016. Diffusion decision model: Current history and issues. *Trends in Cognitive Science*, Vol. 20, pp. 260–281; DOI: 10.1016/j.tics.2016.01.007.

Schwarz, N., & Sudman, S. [Eds.] (1992). *Context Effects in Social and Psychological Research.* Springer Verlag, Berlin.

Shalyt – Margolin, A. (2021). The quantum field theory boundaries applicability and black holes thermodynamics. *International Journal of Theoretical Physics*, Vol. 60, pp. 1858–1869; DOI: 10.1007/s10773-021-04804-1.

Shankar, R. (1994). *Principles of Quantum Mechanics.* Kluwer Academic Publishers, New York.

Sharot, T., Velasquez, C.M., & Dolan, R.J. (2010). Do decisions shape preference? Evidence from blind choice, *Psychological Science*, Vol. 21, No. 9. pp. 1231–1235; DOI: 10.1177/0956797610379235.

Sides, A., Osherson, D., Bonini, N., & Viale, R. (2002). On the reality of the conjunction fallacy. *Memory & Cognition*, Vol. 30, pp. 191–198; DOI: 10.3758/BF03195280.

Slavnov, D.A. (2006). The possibility of reconciling quantum mechanics with classical probability theory. *Theoretical and Mathematical Physics*, Vol. 149, No. 3, pp. 1690–1701; DOI: 10.1007/s11232-006-0151-0

Sloman, S.A. (1996). The empirical case for two systems of reasoning. *Psychological Bulletin*, Vol. 119, pp. 3–22.

Tentori, K., & Crupi, V. (2012). On the conjunction fallacy and the meaning of and, yet again: A reply to Hertwig, Benz, and Krauss (2008). *Cognition*, Vol. 122, pp. 123–134; DOI: 10.1016/j.cognition.2011.09.002.

Trueblood, J.S., & Busemeyer, J.R. (2011). A quantum probability model of casual reasoning. *Frontiers in Cognitive Science*, Vol. 3; DOI: 10.3389/fp-syg.2012.00138.

Trueblood, J.S., & Hemmer, P. (2017). The generalized quantum episodic memory model. *Cognitive Science*, Vol. 41, No. 8, pp. 2089–2125; DOI: 10.1111/cogs.12460.

Trueblood, J.S., Yearsley, J.M., & Pothos, E.M. (2017). A quantum probability framework for human probabilistic inference. *Journal of Experimental Psychology: General*, Vol. 146, No. 9, pp. 1307–1341; DOI: 10.1037/xge0000326.

Tversky, A., & Shafir, E. (1992). Choice under conflict: The dynamics of deferred decision. *Psychological Science*, Vol. 3, No. 6, pp. 358–361; DOI: 10.1111/j.1467–9280.1992.tb00047.x.

Tversky, R., & Kahneman, D. (1974). Judgment under uncertainty: Heuristics and biases. *Science*, Vol. 185, pp. 1124–1130.

Tversky, R., & Kahneman, D. (1983). Extensional versus intuitive reasoning: The conjunctive fallacy in probability judgment. *Psychological Review*, Vol. 90, pp. 293–315.

Tversky, R., & Kahneman, D. (1986). Rational choice and the framing of decisions. *The Journal of Business*, Vol. 59, No. 4, Part 2: The Behavioral Foundations of Economic Theory, pp. 251–278.

Vedral, V. (2006). *Introduction to Quantum Information Science*. Oxford University Press, Oxford.

von Neumann, J. (1955). *Mathematical Foundations of Quantum Mechanics*. Princeton University Press, Princeton.

Waddup, O., Blasiak, P., Yearsley, J.M., Wojciechowski, B.W., & Pothos, E.M. (2021). Sensitivity to Context in Human Interactions. *Mathematics*, Vol. 9, No. 2784; DOI: 10.3390/math9212784.

Waddup, O.J., Yearsley, J.M., Blasiak, P., & Pothos, E.M. (2023). Temporal bell inequalities in cognition. *Psychonomic Bulletin & Review*; DOI: 10.3758/s13423-023-02275-5

Wang, Z., & Busemeyer, J.R. (2013). A quantum question order model supported by empirical tests of an a priori and precise prediction. *Topics in Cognitive* Science, Vol. 5, No. 4, pp. 689–710; Doi: 10.1111/tops.12040.

Wang, Z., Busemeyer, J.R., Atmanspacher, H., & Pothos, E.M. (2013). The potential of using quantum theory to build models of cognition. *Topics in Cognitive Science*, Vol. 5, No. 4, pp. 672–688; DOI: 10.1111/tops.12043.

Wang, Z., Solloway, T., Shiffrin, R.M., & Busemeyer, J.R. (2014). Context effects produced by questions orders reveal quantum nature of human judgment. *PNAS*, Vol. 111, No. 26, pp. 9431–9436; DOI: 10.1073/pnas.1407756111

Wenger, M.J., & Townsend, J.T. [Eds.] (2001). *Computational, Geometric, and Process Perspectives on Facial Cognition: Contexts and Challenges*. Lawrence Erlbaum Associates Publishers, Mahwah.

Wheeler, J.A., & Zurek, W.H.(1983). *Quantum Theory and Measurement*. Princeton University Press, Princeton.

White, L.C., Barque – Duran, A., & Pothos, E.M. (2016). An investigation of a quantum probability model for the constructive effect of affective evaluation. *Philosophical Transactions A*, Vol. 374, No. 20150142; DOI: 10.1098/rsta.2015.0142.

White, L.C., Pothos, E.M., & Busemeyer, J.R. (2014). Sometimes it does hurt to ask: The constructive role of articulating impression. *Cognition*, Vol. 133, pp. 48–64; DOI: 10.1016/j.cognition.2014.05.015.

White, L. C., Pothos, E. M., & Jarrett, M. (2020). The cost of asking: How evaluations bias subsequent judgments. *Decision*, Vol. 7, No. 4, pp. 259–286; DOI: 10.1037/dec0000136.

Wojciechowski, B.W., Izydoryczk, B., Blasiak, P., Yearsley, J., White, L.C., & Pothos, E.M. (2022). Constructive biases in clinical judgment. *Topics in Cognitive Science*, Vol. 14, pp. 508–527; DOI: 10.1111/tops.12547.

Wojciechowski, B.W., & Pothos, E.M. (2018). Is there a conjunction fallacy in legal probabilistic making? *Frontiers in Psychology*, Vol. 9, No. 391; DOI: 10.3389/fpsyg.2018.00391.

Yearsley, J.M., & Busemeyer, J.R. (2016). Quantum cognition and decision theories: A tutorial. *Journal of Mathematical Psychology*, Vol. 74, pp. 99–116; DOI: 10.1016/j.jmp.2015.11.005

Yearsley, J.M., & Pothos, E.M. (2014). Challenging the classical notion of time in cognition: A quantum perspective. *Proceedings of the Royal Society B*, Vol. 281, No. 20133056; DOI: 10.1098/rspb.2013.3056.

Yearsley, J.M., & Pothos, E.M. (2016). Zeno's paradox in decision – making. *Proceedings of the Royal Society B*, Vol. 283; DOI: 10.1098/rspb.2016.0291.

Yukalov, V., & Sornette, D. (2010). Decision theory with prospective interference and entanglement. *Theory and Decision*, Vol. 70, pp. 283–328; DOI: 10.1007/s11238-010-9202-y.

Yukalov, V.I., & Sornette, D. (2014). Conditions for quantum interference in cognitive sciences. *Topics in Cognitive Science*, Vol. 6, pp. 79–90; DOI: 10.1111/tops.12065.

3

CONTEXTUALITY AND ASSESSMENT OF GUILT

Order effects and legal decision-making

The essential premise of the judiciary is that a court's judgement, decree or decision, as well as the interpretation, application and enforcement of existing legislation in relation to a particular set of circumstances are all the result of logical reasoning and rational decision-making. A judge can make multiple critical rulings during the course of a single case. A judge determines whether or not an accused may remain free until trial, whether or not evidence is admissible, and whether or not evidence should be included as a relevant source of information in a case. The length of probation is also decided by the judge. To reach a verdict, judges must classify and analyse (e.g., in terms of how credible it is) evidence and assign a value to the evidence. Judges, together with administrators who set the applicable rules and procedures, have occupation that are quite complex from a cognitive perspective. A portion of their role is to think as applied cognitive psychologists, recognizing the strengths and limitations of the relevant mental functions and how these might enrich and strengthen the criminal justice system.

Some research on legal decision-making has concentrated on external elements such as pre-trial publicity, inadmissible evidence, the science of evidence, racial stereotypes and eyewitness credibility based on attractiveness or other fictitious traits (Spellman & Tenney, 2010). Additional research has focused on memory issues, such as lawyers' inability to determine whether a witness accurately identifies an alleged offender, the reliability of memories, and whether a confession provides an accurate description of a suspect's actions (Howe, 2013; Simon, 2012). Research in simulated and actual legal contexts has also demonstrated biases in court judgements, such as instances

DOI: 10.4324/9781003402480-3

of judges justifying a defendant's acts based on their own personal experiences, preconceptions and biases (Green & Wrightsman, 2003; Saks & Thompson, 2003). It has also been demonstrated that attorneys underestimate probabilistic data and fail to comprehend a range of statistical principles and pertinent methodological concerns (Daftary-Kapur et al., 2010; McAuliff et al., 2003; Spellman & Tenney, 2010). Hence, even with identical evidence, dramatically diverse judgements are conceivable (Green & Wrightsman, 2003).

Order effects accompanying the evaluation of evidence, for instance, are an intriguing form of prejudice. It is well-known, based on study in social psychology, that previous ideas or decisions can establish unique perspectives or settings for evaluating subsequent ones (Schwarz, 2007). In addition, it often appears that material presented first or last is favoured in overall evaluations (Hogarth & Einhorn, 1992). There is evidence that these biases also impair legal decision-making processes. Similar biases have been documented in the legal decision-making process. For instance, Dahl and colleagues (2009) showed that the most recent piece of evidence had a bigger impact on the final assessment of a (hypothetical) suspect's guilt, which is an example of the recency effect. Price and Dahl (2013) hypothesized that such an impact may be a contrast effect if the final piece of information contrasts with and has a different valence than the first. The alternate argument asserts that the concluding piece of evidence is only more accessible.

There is evidence of asymmetric processing or, at the very least, of asymmetric scepticism. For instance, Marksteiner and colleagues (2011) observed that if the initial hypothesis is that a suspect is guilty, incriminating evidence is rated as more trustworthy than exonerating evidence. In a comparable study, however, Maegherman and colleagues (2022) found only recency effects, and not an imbalance in the weighting of information based on whether it suggested guilt or innocence. As participants investigated contradicting materials, the probability of guilt increased when incriminating evidence was offered last and reduced when exonerating evidence was presented last. Charman and colleagues (2015) discovered context effects when initial belief indicated guilt (but not innocence) and when assessing subsequent evidence (but not prior evidence). At least the latter result could be explained by the notion that review is delayed until a decision point and, at that time, more accessible information is favoured (Costabile & Klein, 2005). Similar findings have led to a variety of theoretical suggestions, such as confirmation biases (Kassin et al., 2013), asymmetric scepticism (Ask & Granhag, 2007; Ask et al., 2008), and cognitive coherence frameworks (Holyoak & Simon, 1999; Simon et al., 2001).

Psychological research may add to our understanding of cognitive processes in legal settings and the elements that may impair our expectation of a reasonable and unbiased evaluation of evidence. Importantly, cognitive limitations or biases may impair the accuracy of criminal verdicts and damage the adjudicative process, so that it is ill-equipped to discriminate between truth and error, falling short of the criminal justice system's required level of precision (Simon, 2012). There is a significant body of research on so-called judgement and choice errors in cognition (Hertwig et al., 2013; Kahneman et al., 1982). Yet, many of these biases arise with fabricated decision scenarios in psychology investigations of low significance and debatable participant interest. It is also potentially erroneous to extend an image of "irrational" decision-making from psychology research with modest stakes to legal actions that may have enormous ramifications for the individuals involved (Daftary-Kapur et al., 2010; Spellman & Tenney, 2010).

Quantum probability theory assumes that at any moment a system is in an indefinite superposition state until measurement is performed on the system. To be in a superposed state means that all possible definite states have the potential for being actualized, but only one of them will become actual upon measurement (Bruza et al., 2015). Several enlightening findings demonstrate how a judgement or evaluation might influence subsequent cognitive processing (e.g., later judgements, opinions or preferences). For instance, Wang and colleagues (2013, 2014) empirically examined 70 nationally representative surveys and two laboratory studies that changed question order. Findings gave substantial support for the quantum question (QQ) equality, a straightforward explanation for a startling pattern involving measurement order effects in social and behavioural research, which was originally developed to explain the noncommutativity of measurements in physics. Busemeyer and Wang (2017) conducted the ABA experiment to test a crucial prediction of quantum probability theory when applied to human judgements and decisions. The prediction relates to the impact of incompatible measures on one another (i.e., the answers to the measurements depend on their order). Immediately after the initial measurement of an opinion on topic A, the response to a second measurement on issue A will be equal. But, according to the uncertainty principle, if a measurement of opinion on issue A is followed by an incompatible measurement on issue B, the answer to a second measurement on issue A becomes questionable. This idea was tested with 325 individuals in 12 incompatible scenarios. In contrast to past claims, collected data confirm the prediction of quantum probability theory applied to human judgements.

Order effects were also detected while evaluating evidence in study on diagnosis tasks involving medical trainees (Bergus et al., 1998) and a jury decision-making task (McKenzie et al., 2002; Trueblood & Busemeyer, 2011). Another study showed that making a selection can affect one's preference for that selection. For instance, Sharot and colleagues (2010; see also Ariely & Norton, 2008; Brehm, 1956) showed that a blind choice between two alternatives affected post choice preference for the option. In other research, it has been demonstrated that in tasks dependent on signals or factors, decisions modify the way those cues or variables are viewed such that they are more compatible with the decisions. Gloeckner and colleagues (2010) observed coherence shifts. There are changes in the subjective cue validity associated with a decision, showing increased congruence with the decision. Similarly, in a hypothetical complex court case, Holyoak and Simon (1999) demonstrated that the appraisal of arguments altered to become more congruent with the eventual verdict, and Simon and colleagues (2001) generalized this conclusion across a number of conditions. Yet, choices and judgements are not the only factors that might affect later behaviour. For instance, in affect labelling task, expressing an emotion might reduce its intensity in following comments (Torre & Lieberman, 2018).

White and colleagues (2014, 2016) reported a novel decision bias known as the constructive bias. In a sequence of two stimuli with opposing affective valence, evaluating the first stimulus results in a more divergent judgement of the second stimulus than when only the first stimulus is seen. The constructive bias is consistent with a lengthy history of constructive impacts or choice biases in questionnaire judgements. Using the unique function of evaluations in quantum probability, a quantum probability model was utilized to predict the evaluation bias. In a series of experiments White and colleagues (2014, 2020) examined pairings of stimuli (e.g., commercials for smartphones or celebrity faces) with opposite valence over a number of studies (positive, P, or negative, N; trustworthy or untrustworthy). The difference in the rating of the second stimuli was studied, depending on whether or not the first stimulus was separately rated (double-rating condition) or not, that is if both stimuli were assessed in a single rating. Compared to the single rating condition, the double-rating condition consistently resulted in a greater intensity rating for the second stimulus. In the experiments, affective evaluations task were completed by the participants, but in some cases pairs of stimulus were in a positive and then negative (PN) order, and in other variants in a negative and then positive (NP) order. Results show, that in the PN condition, the rating of the second stimulus was more negative when the first stimulus was also rated, as opposed to when the first stimulus was not assessed. Similarly, in the NP condition, a moderate evaluation of the first stimulus resulted in a more favourable appraisal of the second stimulus when paired with a moderate evaluation of the first stimulus (White et al., 2014, 2016). The evaluation bias is an unexpected finding in light of the traditional theories, as the difference between evaluation and observation for the first stimulus can robustly result in a stronger impression for the second stimulus for the same person responding twice to the same stimulus. The intermediate evaluation modifies rating, even though affective information regarding the initial stimulus might be obtained simply by seeing the initial stimulus.

Evaluation bias and quantum probability theory

Quantum probability representations are built on vector spaces with multiple dimensions known as Hilbert spaces. It is considered that a state vector in a multidimensional space corresponds to the relevant cognitive state. In such spaces, distinct subspaces indicate distinct cognitive state vector possibilities. For instance, there are two potential outcomes for the suspect Smith: whether he is guilty (represented by a one-dimensional subspace; a ray) or not guilty (represented by a second subspace, a ray). A projection is a crucial operation in the quantum operationalization – it "lays down" a vector on a specific subspace. For the system represented by the state vector, the squared length of a projection along a subspace defines the likelihood that the corresponding option is true, according to a fundamental theorem of quantum probability theory. Hence, if we ask a judge or jury if Smith is guilty, the squared length of the projection in the guilty subspace is the probability that the answer will be yes. In addition, the answer to the inquiry alters the position of the state vector; the cognitive state vector becomes a vector (of length one) along the projection, so any following question will be answered from a different position (context).

Cognitive representations of quantum probability can be understood primarily with the idea that a greater projection (or overlap) between the state vector and a subspace indicates a greater likelihood. A greater consistency between two possibilities suggests a smaller angle between the respective subspaces, since it is easier to project from one subspace to the other when the angle is small. Similarly, mutually incompatible alternatives are represented by a 90-degree angle between subspaces, as is the case for the guilty – not guilty subspaces. This is because the projection to the not guilty subspace is zero when the state vector is along the guilty subspace. In other words, when the chance of guilt equals one, the probability of innocence is zero.

In quantum probability theory, dynamic processes can be represented as the rotations (movements) of the state vector. The new information received by the decision-maker cause rotation toward the appropriate subspace, affecting the probability For instance, if the judge obtains information indicating the suspect's guilt (the suspect's fingerprints were found at the crime site), the cognitive state vector will be rotated towards the "guilty" ray, and the probability that the judge will finding the defendant guilty increases. When the defence presents evidence showing that someone else was present at the crime scene, the judge's cognitive state vector will be pushed towards the "innocent" ray, resulting in a lower probability of guilt. The cognitive state vector would be rotated in the direction of the innocent ray, which would enhance the projection on the innocent ray and, thus, the likelihood that Smith will be exonerated when judged increases (Busemeyer et al., 2015; White et al., 2014, 2016, 2020).

Study on evaluation bias in legal settings

Using summaries of criminal legal cases and a sample of judges, prosecutors, defence counsel and layperson, we examined the evaluation bias. Summaries of the criminal cases presented descriptions of pieces of evidence demonstrating the guilt or innocence of accused. We were willing to investigate the possibility of asymmetric primacy/recency effects, depending on whether the first evidence supports guilt or innocence. Since evaluation bias is a potential means by which unethical prosecutors could affect the interpretation of specific pieces of material, it was particularly pertinent to evaluate constructive influence in legal decision-making. There has been little prior research on basic processing biases and mistakes in legal decision-making (Jaeger & Trueblood, 2019; Wojciechowski & Pothos, 2018), but there has never been an analysis of the evaluation bias – the core purpose of the study presented in this chapter.

Concerns have been raised regarding the transferability of choice or memory biases from general psychology to legal settings. Significantly, lawyers and psychologists do not believe it is possible to gain a correct grasp of the legal system and a reliable description of legal decision-making independently of the characteristics of the participants, experimental materials and dependent measures (Kapardis, 2003; Konecni & Ebbesen, 1979). For instance, psychological studies conducted purely with student samples and without the input of legal professionals may misrepresent behaviour in actual court cases (Fox et al., 2011). Legal decision-making and cognitive research should be undertaken under conditions resembling criminal court procedures and using stimuli derived from actual cases, according to this viewpoint. Consequently, the overarching objective of this study was to evaluate putative biases in legal decision-making using as realistic of a sample as feasible (consisting of judges, prosecutors and defence attorneys, as well as some participants without a legal background) and as realistic of materials as possible (based on actual criminal legal cases). We believe it is relevant to investigate whether legal professionals, particularly judges, may be less susceptible to the biases currently under consideration, given that such biases are difficult to reconcile with assumptions of the classical probability theory.

We have followed the paradigm created by White and colleagues (2014, 2016), which restricts the space to a two-dimensional, sample space with various possibilities represented by rays. The cognitive state of an observer in relation to the suspect's guilt can be represented in this space by a vector. There is a set of rays for the probability that the observer is in a solely suspect guilty vs. suspect innocent cognitive state (as determined by the case summary evaluation), as well as a set of rays for the cognitive impact of processing the exonerating or incriminating description. The level proximity of rays shows the level of consistency of the alternatives.

In the innocent–guilty condition, the rotation begins with the exonerating summary (showing the suspect's innocence) since we presume that the only influence on the cognitive state at that point is the processing of the suspect-favourable description. In the innocent–guilty single rating condition, the introduction of the composite innocent–guilty description results in a constant rotation towards the ray denoting guilt. This is the premise that a stimulus does not cause an absolute cognitive response, but rather a fixed shift from the current state toward the criteria for determining the suspect's guilt or innocence. In other words, reading an incriminating summary does not immediately induce a negative disposition in the judge (suspect is guilty). Rather, it modifies the conviction in a negative direction, and the final evaluation will depend on both the degree of change (expressed by the quantum model's rotation) and the original cognitive state. This is an intuitive assumption that also appears in various forms in a variety of psychological theories (e.g., Hogarth & Einhorn, 1992; Laming, 1984; Stewart et al., 2005). After the second part of the case summary, the rater is asked for his or her opinion regarding the suspect's guilt. If the angle between the vector representing the judge's decision and the negative ray (indicating the suspect's guilt) is smaller, the projection onto the guilty ray will be bigger.

Quantum probability theory assumes that the rater is in a superposition regarding his or her cognitive evaluation of the case summaries unless the cognitive state vector is entirely contained within the guilty or innocent ray. Superposition is a quantum probability theory-specific phrase. A superposition situation between suspect guilty verdict and suspect innocent verdict indicates that both outcomes are possible (there is a non-zero probability of making either evaluation). But it also indicates that the state is inconsistent with either evaluation before reaching a conclusion. Once a decision is taken, the cognitive state vector collapses, shifts to the ray corresponding to the subspace corresponding to the choice.

In the double-rating condition with an intermediate rating for the first portion of the description, the rater forces a transition of the cognitive state vector to either the guilty or innocent ray, depending on the summary's content. This is an extra modification to the state vector, which combines with the modification introduced by the second section of the case summary. Under the innocent–guilty condition, for instance, the intermediate rating moves the cognitive state vector closer to the guilty ray, hence increasing the likelihood of a negative rating. Under the innocent–guilty condition, the intermediate judgement makes the second portion of the description appear more incriminating, and vice versa for the guilty–guilty condition.

The quantum probability model's predictions are as follows: in the innocent–guilty order the initial cognitive state and contrast between the second and the first part of the summary will determine how the second part of the summary will be evaluated. Without the intermediate rating, this cognitive state corresponds to the conviction of innocence; therefore, the perception of the aggravating circumstances is contingent on the contrast between the two (White et al., 2014, 2016). Using the intermediate rating, we suppose the innocent portion of the summary is positively assessed, thus the cognitive state corresponds to the suspect's innocence opinion (this is the constructive step in the quantum probability model). The contrast between the first (positive) and second (incriminating) parts of the description will determine how the second (incriminating) section is perceived. This would make it more apparent that the facts offered in the second section of the summary are incriminating, resulting in a lower overall rating (higher probability that suspect is guilty). The quantum probability theory forecast is derived from simple assumptions on the representation of the task's multiple components (that is, the stimuli and the corresponding cognitive evaluations).

Other quantum probability approaches to decision-making are founded on the same fundamental principles of the theory and, in particular, share the prevailing assumption that a decision must involve a corresponding projection of the state vector (e.g., Aerts, 2009; Asano et al., 2011a, Asano et al., 2012; Busemeyer et al., 2011; Busemeyer & Wang, 2018, 2019; Pothos & Busemeyer, 2009; Yukalov & Sornette, 2010, 2014). Notice that, in addition to essential quantum principles, the psychological assumptions embodied by different quantum cognition models vary. For instance, Busemeyer and colleagues (2011) model for the conjunction fallacy (Tversky & Kahneman, 1983) assumes that when evaluating conjunctions of incompatible options, the most probable is examined first. With the current model, we have to make assumptions about the dynamics. In Pothos and Busemeyer's (2009) prisoner's dilemma (Shafir & Tversky, 1992) model, decision-makers are supposed to be motivated by a desire to maximize profit and a cognitive dissonance bias. We believe that in order to build cognitive theory, the underlying psychological assumptions of any mathematical formalism, such as the quantum probability theory must be supplemented with additional assumptions. Hence, there is a need for consistency between various mathematical operations approaches (as there is for quantum probability models; Jones & Love, 2011). In one of the most recent studies, Wojciechowski and colleagues (2022) examined bias in eating disorder clinical diagnosis and tested the White et al.'s (2014, 2016, 2020) quantum model. The partial evidence for an evaluation bias was strong, suggesting that the eating disorder diagnostic test was influenced by constructive effects and the model used in the study produced good fit. This motivates the application of White and colleagues quantum model in the current study (2014, 2015, 2020).

We assumed that evaluations would differ based on presentation order (summary indicating innocence as first – summary indicating guilt as second vs summary indicating guilt as first – summary indicating innocence as second). We also predicted that the ratings of a positively valenced stimulus (suspect's innocence) will differ depending on whether a previous stimulus with the opposite valence (suspect's guilt) was rated or not, and that the ratings of a negatively valenced stimulus will differ if a positively valenced stimulus was previously rated. We assume that the initial cognitive state of the participant is determined by the first stimulus in sequence. Following the intermediate evaluation of the first stimulus, the cognitive state is altered to be either positive or negative. The intermediate evaluation will alter the mental state in a certain way. As a result of the intermediate rating, the second stimulus with the opposite valence is evaluated from a different cognitive state than it would have been without the intermediate rating. As the second stimulus is valence-opposite to the first, the impression produced by the second stimulus is more striking. Without the intermediate rating, the variations between the stimuli concern not only aspects of their quality but also non-judgemental factors; therefore, the affective difference between the first and second stimuli is diminished. This is how the quantum probability model predicts that the intermediate rating will enhance the contrast between the two stimuli; the intermediate rating will increase the contrast between the two stimuli.

The quantum probability theory predicts that unless the cognitive state vector is entirely within the positive or negative affect ray, the observer is in a superposition with regard to his or her evaluation of the criminal accusations. Prior to making a determination, superposition means that the state is inconsistent with either judgement. And once a decision is taken, the cognitive state vector transitions the ray in accordance with the decision's outcome, and decisions always result in the expected transitions.

In the order of guilty suspect–innocent suspect, the contrast between the second section of the case summary and the cognitive state before to its introduction will affect how adversely the second section will be judged. Without the intermediate innocent rating, we anticipate that the second portion of the description will be positively assessed; therefore, the cognitive state shifts to reflect a favourable evaluation. The perception of the ensuing section of the case summary indicating guilt will therefore depend on the contrast between the judgements of innocence and guilt. In the double-rating circumstances with the intermediate innocence rating, the observer induces a change of the cognitive state vector to the positive judgement ray (innocent) or the negative judgement ray (guilty) for the first portion of description (guilty). This is an extra modification to the state vector, which combines with the modification introduced by the second section of the case summary. We hypothesize that when a description indicating innocence is read first, followed by a description indicating guilt, the cognitive state vector will move closer to the guilty ray, increasing the likelihood that the suspect will be rated as guilty in the final rating, and vice versa for the guilty–innocent condition. For the guilty–innocent case summary condition, where a summary indicating innocence is read before a summary indicating the suspect's guilt, an intermediate rating of guilt for the first part of the case summary will result in a more guilty rating for the second part of the case description than if the first statement were not rated. The prognosis for the guilty–innocent condition is inverted, as the intermediate verdict will make the second half of the offence appear more serious. In both instances, it is anticipated that the intermediate evaluation will enhance the disparity in the feeling of guilt between the two portions of the case description, resulting in a more polarized ranking of the first and second parts.

The hypothesis that when making legal decisions, lawyers are in a state of superposition and that legal decisions play a constructive role will be confirmed if there is a statistically significant difference between suspects' guilt judgements in identical cases based on the presence or absence of an intermediate rating. If in identical tasks the intermediate judgement influences the cognitive state in some way, the prediction that measuring internal state can play a positive function will be validated. In addition, the hypothesis that legal decisions are made in the superposition state will be confirmed if there are statistically significant differences in suspects' guilt judgements in identical cases, depending on the order of presentation, i.e., whether subjects read the first or second part of the case description first.

Material and methods

The research was conducted in Poland using Polish court records and Polish participants. As part of a prior study (Wojciechowski & Pothos, 2018), we were granted access to real-life criminal case records from several District Courts and Appellate Courts in Poland, pertaining to offences committed between 2000 and 2015. This is a summary of Wojciechowski and Pothos (2018), which describes in detail how the resources for our research were selected. For each criminal case, we gathered all pertinent data, including interrogation and interview techniques, expert testimony, and verdicts. Based on the following criteria, we first selected 40 examples from 400. Initially, a case has to come from a different court district than the participants'. Secondly, we sought for about equal proportions of guilty and innocent defendants (based on the court's explanation). Thirdly, we ensured that the suspects' honest claims were supported by independent evidence and that their fraudulent assertions were refuted by evidence. Based on two independent rates, we picked cases for which the case summaries gave sufficient information to judge the guilt or innocence of the suspects (an experienced retired prosecutor and an experienced retired attorney, both recruited by the first author). We made no attempt to balance the gender, age and ethnicity of the suspects throughout the 40 selected case transcripts. Among these 40 transcripts, we chose nine situations in which raters had the highest concordance and preliminary assessment indicated that all information pointed to the suspect's guilt (three cases), one half of the description was incriminating, and the second half was exonerating (three cases), and for three of them, the sequence was reversed.

The participants were provided with summaries of all criminal cases to evaluate in their own time. Hence, we had no influence over the location and timing of the examination. Participants were supposed to study all materials and score the suspects' guilt on a scale from one to ten based on the available evidence, with 1 correlating to clearly innocent and ten to definitely guilty. We calculated, based on participant responses, that it took participants between 20 and 50 minutes to read the summaries and assess the instances. Case summaries and rating sheets were in Polish.

Concerning the primary experimental manipulation, the evidence for each suspect was divided into two portions that had to be read sequentially. In each case, we reversed the presentation order of the two portions so that half of the participants viewed them in one order and the other half in the opposite order. After reading both parts, participants were requested to submit an overall rating in every instance. In half of the cases, participants were also required to rate the first part of the evidence; i.e., after reading the first piece of evidence, they had to submit a preliminary judgement of guilt. The double-rating condition was distinguished from the single rating condition by whether or not a rating was provided for the initial portion. Note that in the original approach of White et al. (2014, 2020), the design was entirely within participants, meaning that the same person rated two stimuli in both the double and single ratings conditions (pairs of stimuli would be shown twice, interspersed by many other pairs). Due to the uniqueness and conceptual character of the stimuli, this was not achievable in the present instance. The single versus double-rating condition was therefore between individuals. The experiment used a 2 × 2 x 4 (participant: judge versus prosecutor vs attorney at law vs layperson) mixed design (Figure 3.1). The participant variable was a factor between participants, while the suspect and rating variables were within participant factors.

Participants

40 criminal court judges (22 women, 18 men, aged between 29 and 66 years, M = 42.25; SD = 7.66; with professional experience ranging from 3 to 30 years, M=13.5; SD=6.30); 40 prosecutors and attorneys (20 women, 20 men; aged between 28 and 60 years, M = 40.45; SD = 8,35; with professional experience ranging from 1 to 35 years, M=10.05; SD=8.18); and 40 participants without legal background (21 women, 19 men; every participant was Polish and was recruited in Poland.

Participation in this research was optional. Participants were informed that the purpose of the study was to investigate legal decision-making and the unique characteristics of judges and lawyers in evidence evaluation systems. Prior to the study, participants received a consent form and, following its conclusion, a debriefing form. The participants were not compensated for their participation. Given the study's emphasis on judges and lawyers (including prosecutors), the only viable option for progressing was the professional and personal relationships of the first author (who is a practising attorney). Specifically, issues occur in any agreement that ends in payment to a judge or

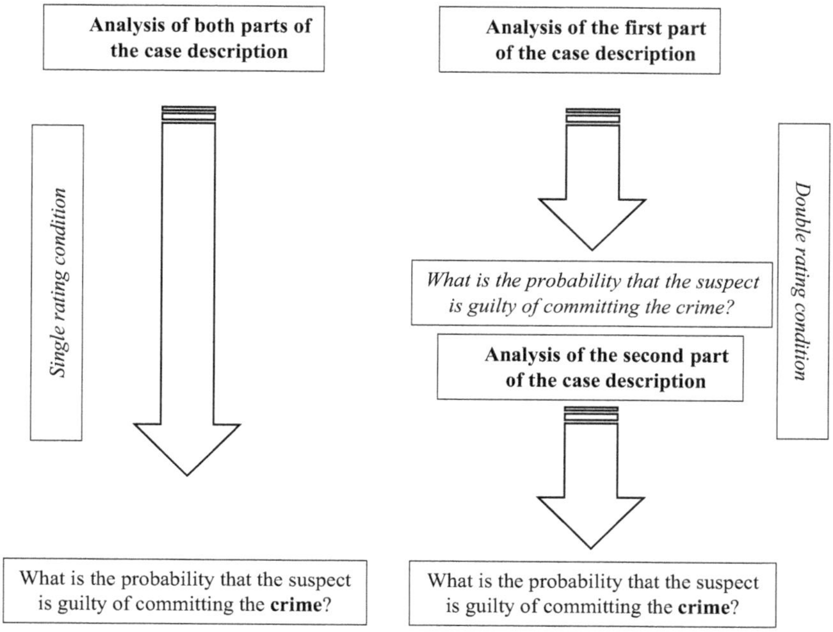

FIGURE 3.1 The design of the study and an outline of the procedure.

prosecutor, and it appears implausible that a judge could be encouraged to participate in a psychology study in exchange for a minor payment. As it was impossible to pay some of the participants, we determined that none of them would be compensated (Wojciechowski & Pothos, 2018). The ethics committee of the Institute of Psychology at the University of Silesia approved the study, and all subjects provided written informed consent prior to participation.

Results

The main purpose of the study was to examine whether there is an evaluation bias and if it depends on the profession (and so relevant expertise) of the participants. In the innocent–guilty condition, the second part indicates culpability of the suspect; we, therefore, expect mean ratings in the double-rating condition to be higher than in the single rating one (recalling that the dependent variable corresponds to degree of guilt's probability). In the guilty–innocent condition, the second part indicates innocence, and so the evaluation bias prediction is that the mean in the double-rating condition would be lower (indicating a lower rating for probability of guilt) than in the single rating one. The guilty–guilty condition was a control one, for which we assume equality in the mean rating for the first and the second parts.

We ran a mixed effects model (multilevel linear model in SPSS with both random and fixed effects) with the final rating of the probability of committing a crime as the dependent variable, three fixed effects, and two random effects. The fixed effects were whether there were judgements just on the second part versus on both parts versus (single vs. double), whether the parts were ordered according to guilty–guilty, innocent–guilty, guilty–innocent (ordering condition), and the rater profession (judge, prosecutor, attorney at law, layperson). The random effects corresponded to the suspect (case) and the participant. The dependent variable was rescaled onto the [0, 1] range, to resemble probabilities for guilt for the judgement following the second part, but we treated the variable as unbounded (i.e., not restricted to the [0, 1] range).

The preliminary analysis showed significant effect for the type of case (guilty–guilty, guilty–innocent and innocent–guilty) with $F(2,12156)=1544.07$, $p<.001$) and assessment method (overall and overall) with $F(1,4194)=3.00$, $p<.05$). The two-way interaction between the single versus double and type of case (guilty–guilty, guilty–innocent, innocent–guilty) fixed effects was also significant ($F(2,6473) = 6.95$, $p<.05$), which is a prerequisite condition for the evaluation bias. The best model for the dependent variable – the final rating of the probability of committing a crime included the fixed effects and their two-way interactions and the random effects were modelled with both intercepts and slopes for the ordering condition (diagonal covariance matrix). We have observed the three-way interaction, involving rater profession ($F(6,1056)=2.63$, $p<.05$). The *post-hoc* procedure revealed differences between suspect's guilt probability ratings depending on type of rating (single overall or double rating). When the case was guilty–guilty, there was no effect of rating for any of the professional groups ($F_{judges}(1,1056)=.59$, $p=.44$; $F_{prosecutors}(1,1056)=.26$, $p=.61$; $F_{attorneys}(1,1056)=.95$, $p=.33$; $F_{layperson}(1,1056)=2.94$, $p=.08$). In the guilty–innocent type of case, there was an effect for prosecutors ($F_{prosecutors}(1,1056)=6.52$, $p<.05$) and layperson ($F_{layperson}(1,1056)=31.71$, $p<.001$) with the overall suspect's guilt probability rating higher, than in the case of intermediate rating, for judges and attorneys there was no main effect ($F_{judges}(1,1056)=1.53$, $p=.22$; $F_{attorneys}(1,1056)=2.07$, $p=.15$). In the innocent–guilty type of case, there was an effect for judges ($F_{judges}(1,1056)=5.79$, $p<.05$) and layperson ($F_{layperson}(1,1056)=14.56$, $p<.001$) with the overall suspect's guilt probability rating significantly lower, than in the case of intermediate rating, for prosecutors and attorneys there was no main effect ($F_{prosecutors}(1,1056)=.03$, $p=.87$; $F_{attorneys}(1,1056)=2.12$, $p=.15$). Detailed comparison of means and standard errors is presented in Table 3.1.

Results provide evidence for constructive effects and the evaluation bias. Probability assessments of the suspects' guilt were influenced by constructive influences – even though summaries of the cases included the same incriminating and exonerating clues, probability of guilt was different in overall and intermediate rating condition. On this basis it can be concluded that raters

TABLE 3.1 Comparison of means and standard errors for the suspects' guilt probability with or without intermediate rating in different groups of participants

	Guilty–Guilty			*Guilty–innocent*			*Innocent–Guilty*		
	With intermediate rating	Overall rating only		With intermediate rating	Overall rating only		With intermediate rating	Overall rating only	
	M	*M*	*SE*	*M*	*M*	*SE*	*M*	*M*	*SE*
Judges	.88	.92	.04	.38	.42	.03	.87*	.69*	.07
Prosecutors	.91	.87	.06	.32*	.45*	.01	.82	.80	.12
Attorneys	.87	.82	.05	.34	.40	.04	.81	.67	.09
Layperson	.92	.84	.04	.32**	.50**	.03	.91**	.63**	.07

*pairwise comparison indicating significant differences with $p<.05$; **pairwise comparison indicating significant differences with $p<.001$

were in an indefinite superposition state, until measurement was performed on the system. All possible definite states – suspects' guilt probability levels – had the potential for being established but became actual only upon measurement.

Moreover, results provide strong support for the White and colleagues (2014, 2016, 2020) quantum model. The main characteristic of this model is that a judgement entails the change of the mental state in a specific way. Based on the quantum model proposed by White and colleagues, we were expecting that when a description indicating innocence will be read first, followed by a description indicating guilt, the cognitive state vector of the participant of the study will move closer to the guilty ray, increasing the likelihood that the suspect will be rated as guilty in the final rating. For the guilty–innocent condition prognosis was inverted, as the intermediate verdict should make the second half of the offence appear more exonerating. In both instances, it was anticipated that the intermediate evaluation will enhance the disparity in the feeling of guilt between the two portions of the case description, resulting in a more polarized ranking of the first and second parts. Additionally, following White and colleagues model we were expecting that in the guilty–guilty scenario there will be no significant difference between intermediate and overall ratings. Results confirmed constructive influence model based on the assumptions of the quantum probability theory. In the innocent–guilty scenario judges and layperson rated probability of suspect's guilt higher when intermediate rating was introduced, and in the guilty–innocent scenario prosecutors and participants without legal background assessed suspects' guilt probability as lower, when the intermediate opinion was expressed. Results show not only that there is a constructive influence, but also that this influence is the same as the one assumed in quantum theory. Study confirmed that state of a rater can be represented in a two-dimensional subspace, with one-dimensional subspaces for guilt and innocence. In the guilty–innocent case, a participant would first be exposed to incriminating information, so the initial mental state vector, ψ, is placed near the subspace for guilt. The angle between ψ and the guilty ray is labelled as *rating*. Introducing the exonerating information results in a rotation, angle *n*, towards the innocent ray. In the single rating condition, the new mental state is ψ_{single}. In the double-rating condition, after the incriminating disorder information, the judgement should first result in a projection of the state vector along the guilty ray. Then, introducing the exonerating information would rotate the state vector along angle *n*, giving us mental state ψ_{double}. In both cases, we compute the probability of innocence by projecting along the innocent ray, and probability of guilt is higher for one rating only and lower when intermediate rating is introduced, which is the evaluation bias.

It is important to note that the current decisions were not designed to be simple to comprehend. Each stimulus was a narrative of a suspect's case, which was believable, fairly detailed, and comprised numerous distinctive qualities. Participation included highly experienced legal experts. Exists evidence that participants in the current tasks behave rationally? According to the normative justification for quantum theory, constructive impacts that are especially consistent with the quantum model are deemed rational in particular circumstances (Pothos et al., 2017). This rationale would not apply to a constructive impact that is contradictory with quantum theory. Since participants' behaviour was governed by quantum rules in regard to positive impacts, they were in theory adhering to logical principles. According to Pothos and colleagues (2017), when the system that is the topic of probabilistic inference is vulnerable to constructive influences, it is sensible to apply quantum concepts.

References

Aerts, D. (2009). Quantum structure in cognition. *Journal of Mathematical Psychology*, Vol. 53, No. 5, pp. 314–348; DOI:10.1016/j.jmp.2009.04.005.

Ariely, D., & Norton, M.I. (2008). How actions create – not just reveal – preferences. *Trends in Cognitive Sciences*, Vol. 12, pp. 13–16; DOI: 10.1016/j.tics.2007.10.008.

Asano, M., Basieva, I., Khrennikov, A., Ohya, M., & Tanaka, Y. (2012). Quantum-like dynamics of decision making. *Physica A*, Vol. 391, pp. 2083–2099; DOI: 10.1016/j.physa.2011.11.042.

Asano, M., Ohya, M., Tanaka, Y., Basieva, I., & Khrennikov, A. (2011a). Quantum-like model of brain's functioning: Decision making from decoherence. *Journal of Theoretical Biology*, Vol. 281, pp. 56–64; DOI: 10.1016/j.jtbi.2011.04.022.

Asano, M., Ohya, M., Tanaka, Y., Khrennikov, A., & Basieva, I. (2011b). On application of Gorini-Kossakowski-Sudarshan-Lindblad equation in cognitive psychology. *Open Systems & Information Dynamics*, Vol. 18, pp. 55–69; DOI: 10.1142/S1230161211000042.

Ask, K., & Granhag, P.A. (2007). Motivational bias in criminal investigators' judgments of witness reliability. *Journal of Applied Social Psychology*, Vol. 37, pp. 561–591; DOI: 10.1111/j.1559-1816.2007.00175.x.

Ask, K., Rebelius, A., & Granhag, P.A. (2008). The 'elasticity' of criminal evidence: A moderator of investigator bias. *Applied Cognitive Psychology*, Vol. 22, No. 9, pp. 1245–1259; DOI: 10.1002/acp.1432.

Bergus, G.R., Chapman, G.B., Levy, B.T., Ely, J.W., & Oppliger, R.A. (1998). Clinical diagnosis and order information. *Medical Decision Making*, Vol. 18, pp. 412–417; DOI: 10.1177/0272989X9801800409.

Brehm, J.W. (1956). Post-decision changes in the desirability of choice alternatives. *Journal of Abnormal and Social Psychology*, Vol. 52, pp. 384–389; DOI: 10.1037/h0041006.

Bruza, P.D., Wang, Z., & Busemeyer, J. R. (2015). Quantum cognition: A new theoretical approach to psychology. *Trends in Cognitive Sciences*, Vol. 19, No. 7, pp. 383–393; DOI: 10.1016/j.tics.2015.05.001.

Busemeyer, J.R., Pothos, E.M., Franco, R., & Trueblood, J.S. (2011). A quantum theoretical explanation for probability judgment errors. *Psychological Review*, Vol. 118, No. 2, pp. 193–218; DOI: 10.1037/a0022542.

Busemeyer, J.R., & Wang Z. (2017). Is there a problem with quantum models of psychological measurements? *PLoS ONE*, Vol. 12, No. 11; DOI: 10.1371/journal.pone.0187733.

Busemeyer, J.R., & Wang, Z. (2018). Hilbert space multidimensional theory. *Psychological Review*, Vol. 125, No. 4, pp. 572–591; DOI: 10.1037/rev0000106.

Busemeyer, J.R., & Wang, Z. (2019). Introduction to Hilbert Space multi – dimensional modeling, (in:) *Quantum-Like Models for Information Retrieval and Decision – Making, STEAM – H: Science, Technology, Engineering, Agriculture, Mathematics & Health*, [eds.] Aerts, D., Khrennikov, A., Melucci, M., Toni, B., Springer, Cham, pp. 41–49; DOI: 10.1007/978-3-030-25913-6_3

Charman, S.D., Carbone, J., Kekessie, S., & Villalba, D.K. (2015). Evidence evaluation and evidence integration in legal decision – making: order of evidence presentation as a moderator of context effects. *Applied Cognitive Psychology*, Vol. 30, No. 2, pp. 214–225; DOI: 10.1002/acp.3181.

Costabile, K.A., & Klein, S.B. (2005). Finishing strong: Recency effects in juror judgments. *Basic and Applied Social Psychology*, Vol. 27, No. 1, pp. 47–58; DOI:10.1207/s15324834basp2701_5.

Daftary-Kapur, T., Dumas, R., & Penrod, S.D. (2010). Jury decision-making biases and methods to counter them. *Legal and Criminological Psychology*, Vol. 15, pp. 133–154.

Dahl, L.C., Brimacombe, C.A., & Lindsay, D.S. (2009). Investigating investigators: How presentation order influences participant-investigators' interpretations of eyewitness identification and alibi evidence. *Law and Human Behavior*, Vol. 33, No. 5, pp. 368–380; DOI: 10.1007/s10979-008-9151-y.

Fox, P., Wingrove, T., & Pfeifer, C. (2011). A Comparison of Students' and Jury Panelists' decision-making in split recovery cases. *Behavioral Sciences and the Law*, Vol. 29, No. 3, pp. 358–375; DOI: 10.1002/bsl.968.

Gloeckner, A., Betsch, T., & Schindler, N. (2010). Coherence shifts in probabilistic inference tasks. *Journal of Behavioral Decision Making*, Vol. 23, No. 5, pp. 439–462; DOI: 10.1002/bdm.668.

Green, E., & Wrightsman, L. (2003). Decision making by Juries and Judges: International perspective, [in:] *Handbook of Psychology in Legal Contexts*, [eds.] Carson, D., & Bull, R., John Wiley and Sons, Chichester, pp. 401–422.

Hertwig, R., & Hoffrage, U., the ABC Research Group. (2013). *Simple Heuristics in a Social World*. Oxford University Press, New York.

Hogarth, R.M., & Einhorn, H.J. (1992). Order effects in belief updating: The belief-adjustment model. *Cognitive Psychology*, Vol. 24, pp. 1–55; DOI: 10.1016/0010–0285(92)90002-J.

Holyoak, K.J., & Simon, D. (1999). Bidirectional reasoning in decision making by constraint satisfaction. *Journal of Experimental Psychology: General*, Vol. 128, No. 1, pp. 3–31; DOI: 10.1037/0096–3445.128.1.3.

Howe, M.L. (2013). Memory development: Implications for adults recalling childhood experiences in the courtroom. *Nature Reviews Neuroscience*, Vol. 14, No. 12, pp. 869–876; DOI:10.1038/nrn3627.

Jaeger, Ch.B., & Trueblood, J.S. (2019). Thinking quantum: A new perspective on decision making in law. *Florida State University Law Review*, Vol. 46, No. 4, pp. 733–806.

Jones, M., & Love, B.C. (2011). Bayesian fundamentalism or enlightenment? On the explanatory status and theoretical contributions of Bayesian models of cognition. *Behavioral and Brain Sciences*, Vol. 34, pp. 169–231; DOI: 10.1017/S0140525X10003134.

Kahneman, D., Slovic, P., & Tversky, A. (1982). *Judgment Under Uncertainty: Heuristics and Biases*. Cambridge University Press, New York.

Kapardis, A. (2003). *Psychology and Law. A Critical Introduction*. Cambridge University Press, Cambridge.

Kassin, S.M., Dror, I.E., & Kukucka, J. (2013). The forensic confirmation bias: Problems, perspectives, and proposed solutions. *Journal of Applied Research in Memory and Cognition*, Vol. 2, No. 1, pp. 42–52; DOI: 10.1016/j.jarmac.2013.01.001.

Konecni, V.J., & Ebbesen, E.B. (1979). External validity of research in legal psychology. *Law and Human Behavior*, Vol. 3, No. 1/2, pp. 39–70.

Laming, D.R.J. (1984). The relativity of "absolute" judgments. *British Journal of Mathematical and Statistical Psychology*, Vol. 37, pp. 152–183; DOI: 10.1111/j.2044–8317.1984.tb00798.x.

Maegherman, E., Ask, K., Horselenberg, R., & van Koppen, P. (2022). Law and order effects: On cognitive dissonance and belief perseverance. *Psychiatry, Psychology and Law*, Vol. 29, No. 1, pp. 33–52; DOI: 10.1080/13218719.2020.1855268.

Marksteiner, T., Ask, K., Reinhard, M.A., & Granhag, P.A. (2011). Asymmetrical scepticism towards criminal evidence: The role of goal- and belief-consistency. *Applied Cognitive Psychology*, Vol. 25, pp. 541–547; DOI: 10.1002/acp.1719.

McAuliff, B.D., Nemeth, R.J., Bornstein, B.H., & Penrod, S.D. (2003). Juror Decision - making in the twenty - first century: Confronting science and technology in court, [in:] *Handbook of Psychology in Legal Contexts*, [eds.] Carson, D., & Bull, R., John Wiley and Sons, Chichester, pp. 303–372.

McKenzie, C.R.M., Lee, S.M., & Chen, K.K. (2002). When negative evidence increases confidence: change in belief after hearing two sides of a dispute. *Journal of Behavioral Decision Making*, Vol. 15, No. 1, pp. 1–18; DOI:10.1002/bdm.400.

Pothos, E.M., & Busemeyer, J.R. (2009). A quantum probability explanation for violations of "rational" decision theory. *Proceedings of the Royal Society B: Biological Sciences*, Vol. 276, No. 1665, pp. 2171–2178; DOI: 10.1098/rspb.2009.0121.

Pothos, E.M., Busemeyer, J.R., Shiffrin, R.M., & Yearsley, J.M. (2017). The rational status of quantum cognition. *Journal of Experimental Psychology: General*, Vol. 146, pp. 968–987; DOI: 10.1037/xge0000312.

Price, H.L., & Dahl, L.C. (2013). Order and strength matter for evaluation of alibi and eyewitness evidence. *Applied Cognitive Psychology*, Vol. 28, pp. 143–150; DOI: 10.1002/acp.2983.

Saks, M.J., & Thompson, W.C. (2003). Assessing evidence: Proving facts, [in:] *Handbook of Psychology in Legal Contexts*, [eds.] Carson, D., & Bull, R., John Wiley and Sons, Chichester, pp. 329–345.

Schwarz, N. (2007). Attitude construction: Evaluation in context. *Social Cognition*, Vol. 25, No. 5, pp. 638–656; DOI: 10.1521/soco.2007.25.5.638

Shafir, E., & Tversky, A. (1992). Thinking through uncertainty: Nonconsequential reasoning and choice. *Cognitive Psychology*, Vol. 24, pp. 449–474.

Sharot, T., Velasquez, C.M., & Dolan, R.J. (2010). Do decisions shape preference?: Evidence from blind choice, *Psychological Science*, Vol. 21, No. 9. pp. 1231–1235; DOI: 10.1177/0956797610379235.

Simon, D. (2012). *In Doubt. The Psychology of the Criminal Justice Process.* Harvard University Press, Cambridge, MA, London, England.

Simon, D., Pham, L.B., Le, Q.A., & Holyoak, K.J. (2001). The emergence of coherence over the course of decision making. *Journal of Experimental Psychology: Learning, Memory, and Cognition*, Vol. 27, No. 5, pp. 1250–1260; DOI: 10.1037/0278-7393.27.5.1250.

Spellman, B.A., & Tenney, E.R. (2010). Credible testimony in and out of court. *Psychonomic Bulletin & Review*, Vol. 17, No. 2, pp. 168–173; DOI: 10.3758/PBR/17.2.168.

Stewart, N., Brown, G.D.A., & Chater, N. (2005). Absolute identification by relative judgment. *Psychological Review*, Vol. 112, No. 4, pp. 881–911; DOI: 10.1037/0033-295X.112.4.881.

Torre, J.B., & Lieberman, M.D. (2018). Putting feelings into words: Affect labeling as implicit emotion regulation. *Emotion Review*, Vol. 10, No. 2, pp. 116–124; DOI: 10.1177/1754073917742706.

Trueblood, J.S., & Busemeyer, J.R. (2011). A quantum probability account of order effects in inference. *Cognitive Science*, Vol. 11, pp. 1518–1552; DOI: 10.1111/j.1551-6709.2011.01197.x.

Tversky, R., & Kahneman, D. (1983). Extensional versus intuitive reasoning: The conjunctive fallacy in probability judgment. *Psychological Review*, Vol. 90, pp. 293–315.

Wang, Z., & Busemeyer, J.R. (2013). A quantum question order model supported by empirical tests of an a priori and precise prediction. *Topics in Cognitive Science*, Vol. 5, No. 4, pp. 689–710; DOI: 10.1111/tops.12040

Wang, Z., Solloway, T., Shiffrin, R.M., & Busemeyer, J.R. (2014). Context effects produced by questions orders reveal quantum nature of human judgment. *PNAS*, Vol. 111, No. 26, pp. 9431–9436; DOI: 10.1073/pnas.1407756111

White, L.C., Barque – Duran, A., & Pothos, E.M. (2016). An investigation of a quantum probability model for the constructive effect of affective evaluation. *Philosophical Transactions A*, Vol. 374, No. 20150142; DOI: 10.1098/rsta.2015.0142.

White, L.C., Pothos, E.M., & Busemeyer, J.R. (2014). Sometimes it does hurt to ask: The constructive role of articulating impressions. *Cognition*, Vol. 133, pp. 48–64; DOI: 10.1016/j.cognition.2014.05.015.

White, L.C., Pothos, E.M., & Jarrett, M. (2020). The cost of asking: How evaluations bias subsequent judgments. *Decision*, Vol. 7, No. 4, pp. 259–286; DOI: 10.1037/dec0000136.

Wojciechowski, B.W., Izydoryczk, B., Blasiak, P., Yearsley, J., White, L.C., & Pothos, E.M. (2022). Constructive biases in clinical judgment. *Topics in Cognitive Science*, Vol. 14, pp. 508–527; DOI: 10.1111/tops.12547.

Wojciechowski, B.W., & Pothos, E.M. (2018). Is there a conjunction fallacy in legal probabilistic making? *Frontiers in Psychology*, Vol. 9, No. 391; DOI: 10.3389/fpsyg.2018.00391.

Yukalov, V., & Sornette, D. (2010). Decision theory with prospective interference and entanglement. *Theory and Decision*, Vol. 70, pp. 283–328; DOI: 10.1007/s11238-010-9202-y.

Yukalov, V.I., & Sornette, D. (2014). Conditions for quantum interference in cognitive sciences. *Topics in Cognitive Science*, Vol. 6, pp. 79–90; DOI: 10.1111/tops.12065.

4

QUANTUM ENTANGLEMENT AND JUDGEMENT ON MULTIPLE OFFENCES

Legal assessment and conjunction fallacy

A key assumption of the judicial system is that a court's judgement, decree or decision, as well as the interpretation, application, or enforcement of an existing law in relation to a specific set of facts, are all the result of rational, accurate decision-making. In the judicial system, such expectations for a reasonable standard in decision-making increase with seniority. In the course of a single case, a judge may be required to make numerous decisions, each of which is crucial to the entire procedures and legal outcome. A judge determines whether an accused remains free pending trial, whether evidence is admissible, and whether sources of information can be used in connection with a crime. Judges frequently determine whether an individual should be found guilty and condemned, or placed on probation, and for how long. In order to render a verdict, judges must analyse evidence in accordance with specific (sometimes extremely complex) standards, classify evidence and apply legal rules. Even though the corresponding expectations for normative decisions are not as high for attorneys and prosecutors as they are for judges, they must also have a sufficient level of familiarity with legal proceedings. However, even lay individuals are sometimes expected to operate at such a high level in legal proceedings (e.g., when members of a jury). To achieve the normative level expected for legal decision-making, it is crucial to understand the psychological strengths and limitations of the human mind. If human intervention in the judicial process can compromise the accuracy of criminal verdicts, then the process may be ill-equipped to discern between truth and error and fall short of meeting the epistemic requirements of the criminal justice system and the certitude it purports to express (Simon, 2012). Therefore, it is vital to investigate and

DOI: 10.4324/9781003402480-4

comprehend the cognitive processes involved in the appraisal of findings and evidence, probability assessment, problem-solving, biases related with legal decision-making, and how and when legal proceedings participants rely on the views of others (Daftary-Kapur et al., 2010; Spellman & Tenney, 2010).

The majority of prior research on legal decision-making has been focused on the influence of external factors, such as pre-trial publicity, inadmissible evidence, scientific evidence, racial stereotypes, the confidence of eyewitnesses or their attractiveness, and confidence in the quality of their knowledge (Spellman & Tenney, 2010; Tenney et al., 2011). Studies indicate that attorneys and prosecutors frequently struggle to determine whether a witness accurately recognised the alleged culprit, whether memories are reliable, and whether a suspect's confession is a credible description of his or her actions (Simon, 2012). Research in simulated and actual legal situations demonstrates that an effective model of judicial decision-making must incorporate the role of human bias (Green & Wrightsman, 2003). For instance, judges prefer to explain a defendant's behaviour in light of their own past experiences and held presumptions (Saks & Thompson, 2003). Similarly, there is evidence that attorneys and prosecutors underuse probabilistic data and fail to comprehend a broad range of statistical principles and methodological concerns (Daftary – Kapur et al., 2010; McAuliff et al., 2003; Spellman & Tenney, 2010). Due to the influence of human biases, attorneys and prosecutors provided with identical evidence may reach vastly different conclusions (Green & Wrightsman, 2003). It has been discovered, for instance, that concluding bits of evidence have far-reaching effects, that the party who presents their argument second has a significant advantage, and that evidence presented at the conclusion of a presentation is favoured (Walker et al., 1972). In a series of studies, Furnham (1986) duplicated the recency effect in the formulation of innocence and guilt judgements of a defendant in a real trial. Similar information acquired in a different order resulted in considerably different final judgements, and the impact extended over time and population, i.e., different participants produced significantly different outcomes. Such results are generally consistent with research across a wide range of specialised decision-making domains, indicating that decisions are susceptible to framing or order effects and other seemingly incidental characteristics of the supplied data and evaluation process (e.g., Hogarth & Einhorn, 1992).

Before attributing a pattern of human decision-making to a non-normative bias, it is necessary to have a criterion for normative decision-making. Typically, psychologists embrace classical probability theory as the gold standard for normative behaviour (Oaksford & Chater, 2009; Tenenbaum et al., 2011). The normative case of the classical probability theory is supported by important mathematical conclusions, including the Dutch book theorem, which states that classically consistent probabilistic assignment protects against certain loss (de Finetti et al., 1993; Pothos et al., 2017). Classical probability theory is applicable whenever an individual must change subjective probabilities in light of new or existing evidence. Several excellent cognitive models based on the classical theory provide solid evidence that humans are predominantly rational (Oaksford & Chater, 2009).

However, there have been recurring differences between the prescription of the classical probability theory and human conduct, and these results raise doubt on any significant claims regarding the rationality of humans in decision-making. Such outcomes are frequently referred to as fallacies to indicate that the corresponding decision-making is flawed. Tversky and Kahneman (1983) demonstrated that conjunctive and disjunctive probability judgement errors are quite robust and can be observed with a wide range of cases. The conjunction fallacy, identified by Tversky and Kahneman (1983) in one of the most prominent studies on decision-making, is one of the most well-known fallacies. Tversky and Kahneman provided participants with brief descriptions of fictitious individuals. In one of their cases, Linda was shown as a smart and opinionated individual. After reading the narrative, participants were asked to rank a series of propositions about Linda according to their perceived likelihood. The essential statements were that Linda is a feminist (F; very probable given the description), a bank teller (BT; highly improbable given the description), and both (a feminist and a bank teller). Tversky and Kahneman's findings suggested that according to participants, probability of Linda being a bank teller and a feminist is higher than the probability of Linda being just the bank teller. The finding that probability of possessing both qualities (bank teller and feminist) is known as the critical finding since it is not possible in the classical probability theory. Conjunction fallacy is problematic for evaluating human decision-making rationality, because formally it is not possible that the probability of a person having both qualities is higher, than having one of them. The conjunction fallacy has been thoroughly repeated, in both Tversky and Kahneman's original study and later work: the fallacy is obviously not prevalent in decision-making, but it is straightforward to establish conditions that reliably lead to the conjunction fallacy (Busemeyer et al., 2011). One could also be inclined to attribute the fallacy to a misunderstanding between the experimenter's intentions and the participant's comprehension. Perhaps, upon seeing the bank teller statement, participants comprehend bank teller and feminist, another conjunction that can be compared with greater flexibility to both. There has been extensive research conducted on this topic.

Despite the fact that various disambiguation techniques can lessen the frequency of conjunction fallacy, residual effect persist (e.g., Dulany & Hilton, 1991; Moro, 2009; Sides et al., 2002; Tentori & Crupi, 2012; Tentori et al., 2013). Importantly, research shows that conjunction fallacy is not simply a product of a lack of motivation. Even if real dollars are at stake, and people have opportunity to gamble and win more money, irrational decisions with betting are made and conjunction fallacy occurs (Bar-Hillel & Neter, 1993;

Sides et al., 2002). Even more crucial is a study conducted by Guthrie and colleagues (2009) showing that not only undergraduate students are prone to the conjunction fallacy, but it can also be observed in federal judges.

The early proponents of the conjunction fallacy regarded the conclusion as suggesting that, in at least some instances, human decision-making is governed by principles other than the classical probability theory, specifically individual heuristics and biases (Pothos & Yearsley, 2023; Tversky & Kahneman, 1983). Numerous attempts have been made to explain the fallacy, even discounting concerns over how participants interpret the key claims. When participants evaluate the conjunction, they are in a different mental state or context than when they evaluate the conjunct. The conjunction result can thus be expressed as probability of Linda having both qualities in context A, which can be higher than probability of Linda being only bank teller in different context (B), and the conditionalizing variable no longer entails a fallacy. However, the majority of decision scientists are not satisfied with such explanations. Instead, it is common to suppose that there are two distinct decision-making processes: one that is fully normative and yields conclusions according with the principles of classical probability theory, and one that is rapid and approximative, for quick, somewhat insignificant decisions (DeNeys & Goel, 2011; Elqayam & Evans, 2013; Kahneman, 2001; Sloman, 1996). Given the importance of decision-making in legal situations to our lives, such an approach would suggest that if humans are capable of normative decision-making at all, it would be evident in such circumstances; but, as we have seen, decision errors can also arise in such contexts (Saks & Thompson, 2003). Conjunction error contradicts classical probability theory, which is why it is known as fallacy. However, a simple yet general model for these types of discoveries, based on the quantum probability theory, has been developed. Models of decision-making grounded in quantum probability theory produce naturally many of the phenomena, such as heuristics and biases, including, but not limited to, conjunction fallacy (Jaeger & Trueblood, 2019; Narens, 2014; Pothos et al., 2015).

Conjunction fallacy and quantum probability theory

Why would someone ever consider using this quantum-based strategy to model the conjunction fallacy? First, classical probability theory does not appear to provide a natural way to comprehend the conjunction error (but note that there are more elaborate classical probability theory models, which we consider below). Second, the heuristic explanations offered for the original conjunction error, such as the similarity between Linda and her numerous probable features (Tversky & Kahneman, 1983, 1986), are intuitive, but imprecise and have limited predictive power. Third, given that the conjunction fallacy problem involves probabilistic reasoning, we may be predisposed to create a model for the conjunction fallacy based on probability theory. What is a probability theory consistent with the intuitions provided in heuristic, similarity-based models for the conjunction fallacy that is not limited in the same manner as classical probability theory? Quantum probability theory is the most immediate response (Bruza et al., 2015; Pothos & Busemeyer, 2022).

Quantum probability theory is the set of principles for assigning probabilities to events in quantum mechanics, excluding the physics. Quantum theory is an alternative to the classical probability theory that gives a set of rules for probabilistic inference (Broekaert et al., 2017; Korf, 2015; Pothos et. al, 2013, 2015; Pothos & Trueblood, 2015). Exact empirical results problematic from a classical perspective, such as the conjunction fallacy, have motivated the pursuit of quantum cognitive models. They have been utilised to accommodate a variety of choice errors, such as order effects in evidence evaluation and legal decision-making (Aerts & Aerts, 1995; Busemeyer & Bruza, 2011; Busemeyer et al., 2011; Haven & Khrennikov, 2013; Pothos & Busemeyer, 2009, 2013; Pothos & Trueblood, 2015; Trueblood & Busemeyer, 2010, 2011; Wang et al., 2014; White et al., 2014, 2016). Quantum probability theory inference is highly context- and perspective-dependent, such that the same question may have different meanings depending on whether it is posed alone or in conjunction with other questions (Ariely & Norton, 2008; Yukalov & Sornette, 2014). Some questions in the quantum probability theory are assumed to be compatible, and the probabilistic picture for such questions is virtually identical to the classical situation. Other questions, however, are incompatible, and incompatible questions are prone to significant contextual and order effects.

Modelling of the conjunction fallacy in quantum probability theory, according to the model proposed by Busemeyer and colleagues (Busemeyer et al., 2015; Pothos & Busemeyer, 2022), is based on four assumptions. The first assumption is that after reading the story about Linda, an individual generates an initial belief state, that represents the individual's views regarding traits or properties that may or may not be true about Linda. This belief state is formally a vector in a multi-dimensional vector space. This belief state is used to respond to any possible question regarding Linda. The second assumption is that questions like "Is Linda a feminist?" are represented in subspaces of a multi-dimensional vector space. There is a subspace for each question, so "Is Linda a bank teller?" has representation in a different subspace, other dimension than question "Is Linda a feminist?" Answers to the questions are projections of the state vector onto the subspaces representing questions. The third assumption is that the projectors do not commute, and therefore, the sequence in which they are applied is significant. The following explains why these two projectors are incompatible. The two notions (feminist and bank teller) are rarely experienced together; therefore, the individual has not created a compatible representation of beliefs regarding combinations of both concepts based on a common experience. A person may have created one basis to represent characteristics of feminists, but this basis is distinct from the basis used to represent characteristics of bank tellers. The concepts do not share the same foundation and are therefore incompatible, requiring the personnel to progressively switch from one basis to the next in order to answer questions concerning each concept. The fourth assumption involves the order in which the concepts are processed in response to the question, "Is Linda a feminist and a teller?" Given that the events are incompatible, the individual must select a processing sequence. Assume that the more probable occurrence will be processed first. It is relatively simple to determine the order of every single question ("Is Linda a feminist?" and "Is Linda a bank teller?). However, the question regarding "feminist and a bank teller" is more nuanced and not as straightforward as the preceding two. The evaluation of the combination involves an additional and more nuanced evaluation of the conditional probability, since it is rating to the question if Linda is a bank teller, given that that she is a feminist. These four assumptions are applied to calculate the quantum predictions for bank teller probability (Busemeyer et al., 2015).

We may explain a quantum probability theory technique for modelling conjunction fallacy in legal decision-making by analogy to Busemeyer's and colleagues (2011) quantum model for the conventional conjunction fallacy. The illustration is based on a considerably simplified sketch of the complete model, yet this suffices for the reasons at hand. The initial phase of the quantum probability theory is to define a vector space in which each subspace corresponds to the outcomes of distinct questions. In the simplest situation, these subspaces have a dimension of one (rays). In the second stage, a (normalised) vector representing the mental state just prior to answering the questions is determined. Note that in the quantum probability theory, probabilistic evaluation is always (mental) state dependant, whereas in the classical probability theory, the role of the mental state, if any, must be seen as conditionalizing information. The third step is to compute probabilities using the quantum-like rules. Probability is computed as the length squared of the projection, which is determined by a projector operator applying a vector to a subspace. In the case of the classic conjunction fallacy, an illustrative caricature of the model is depicted in Figure 4.1, which was created based on the minimal assumptions that the initial mental state must be closer to the answer "yes" to the question regarding Linda's feminism and to the reply "yes" to the question referring to her profession. A conjunction for incompatible questions in the quantum probability theory can only be computed sequentially, since the truth/falsity of incompatible questions cannot be determined concurrently. It is evident that it is possible to acquire the conjunction fallacy as a "right" probabilistic outcome. From the initial perspective of the Linda story, it is quite difficult to see Linda as a bank teller, but once we accept Linda as a feminist, it becomes simpler to imagine her as a bank teller as well.

Figure 4.1 presents a simple two-dimensional space to explain how the conjunction fallacy can be modelled (please note that the model's specification is not constrained to one-dimensional subspaces). The probability for the bank teller alone inquiry is derived from the direct projection of the initial state Ψ to the bank teller (BT) axis, which is represented by the shorter light-grey vertical segment. The likelihood for the conjunction is depicted by first projecting Ψ onto feminist (F) axis and then onto bank teller axis, which is represented by the longer dark grey vertical segment. Note that the projection for the conjunction (vertical section in dark grey) exceeds the projection for the bank teller alone (light-grey vertical segment). Same model is also applicable in other contexts, such as determining the likelihood that a person has committed two offences.

Model presented in this section can account for conjunction fallacy, disjunction fallacy, averaging effects and order effects. Using quantum concepts to create models of probability decisions is, however, merely one of numerous conceivable approaches. It is worth to add that Aerts (2009) and his colleagues (Aerts & Gabora, 2005) have proposed alternative quantum models that account for conjunction and disjunction mistakes in conceptual combinations. Importantly, their model may generate double conjunction mistakes; however, it requires parameter adjustments to account for the distinctions between conjunction and disjunction errors.

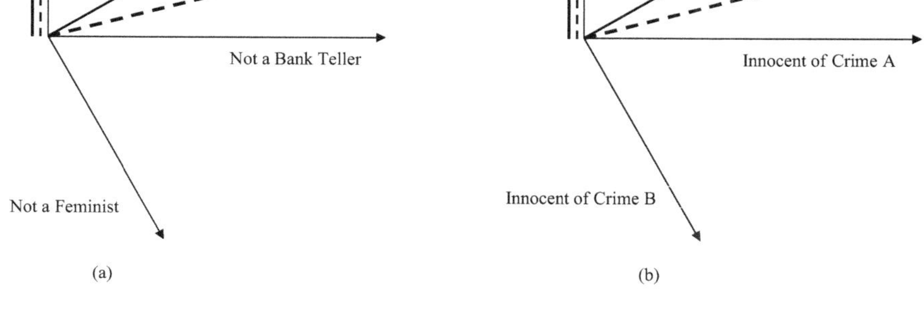

FIGURE 4.1 An illustration of the quantum probability (QP) approach for the (single) conjunction fallacy in (a) the classic "Linda problem" and in (b) legal decision-making.

Study on conjunction fallacy and legal decision-making

With the aforementioned considerations in mind, we have conducted a study to examine whether the conjunction fallacy can be observed in legal decision-making and to show how quantum probability theory can be used to explain experimental findings. Study has been conducted using participants with a wide range of legal expertise, from (legally) naive individuals, attorneys and prosecutors, to judges. From the latter category of professionals, the highest possible standard of legal decision-making can be expected and, if there is truth to dual-route models of human decision-making, they should incorporate the conjunction fallacy.

The experimental inquiry is based on the presentation of information regarding two separate crimes that probably were committed by the same suspect. Each crime's detail is given sequentially. Some participants had to individually determine whether suspects committed each of the two offence (expressing two opinions), while others had to determine if the suspect was guilty of both crimes in a single conjunctive judgement. Traditionally, if one were to consider all 100 of these suspects, some of them would have only done the first crime, while others would have committed the second. For a suspect to have done both crimes, he or she must have also committed one – the likelihood of the conjunction (committing both crimes) must be smaller than the chance of either (committing each crime individually). However, psychologically, the logic of the design is that if a person is, for instance, found guilty of one crime, he/she is likely to have committed the other crime as well, particularly in cases of related crimes, which is indeed close to one of Tversky and Kahneman's (1983) assumptions for when to expect a conjunction fallacy, that is when there is some causal relationship between the conjuncts.

Material and methods

During earlier research, we have obtained access to real-life criminal case records from several District Courts and Appellate Courts in Poland pertaining to offences committed between 2000 and 2015. Each criminal case's papers, including interrogation and interview techniques, expert testimony, and rulings, were thoroughly scrutinised and photocopied. We initially picked 60 of these 400 criminal cases based on the following criteria. First, we selected court records from different court districts than those of the (legal profession) participants to reduce the likelihood that they would evaluate cases they had previously encountered. Second, we searched for a balance between circumstances in which the suspect was guilty and cases in which the suspect was innocent, namely cases in which the suspect was innocent of both offences and cases in which the suspect was guilty of both. In specific circumstances, the suspect's guilt was proven on the basis of the ruling court's justification. In circumstances of truthful statements from the suspect, we required overwhelming evidence independent of the suspect's statement to corroborate the statement, and in cases of false statements, overwhelming evidence outside of the suspect's statement to disprove it. These inclusion criteria are appropriate considering that, on the one hand, whole truths are typically unavailable in criminal cases and, on the other hand, precise classification of the transcripts was essential for the validity of the study. Fourth, while the study was based on case summaries (using full records would have been unfeasible), we confirmed that the case summaries accurately represented the suspect's guilt or innocence. This was determined by having two independent, competent raters – an experienced retired prosecutor and an experienced retired attorney recruited by the experimenter – examine each summary and determine if it led to a conclusion regarding the suspect's guilt or innocence for one or both offences that was consistent with the trial outcome. In the main experiment, only case summaries with confirmed valence from this early stage were utilised. Factors such as the type of offence, the suspect's gender, age and race varied between transcripts (but could not be fully counterbalanced). Using the aforementioned criteria, we were able to discover 18 case summaries (out of the original 60 instances), categorised as follows: nine summaries of cases in which the suspect was found guilty of both counts, and nine summaries of cases in which the suspect was found innocent of both accusations.

According to the regulations governing Polish criminal proceedings, the judge actively participates in the examination of the evidence. All courts are presided over by judges selected by the President of the Republic based on the advice of the National Council of the Judiciary. In Poland, judges are independent and exclusively answerable to the law. Minimum age for employment is 26, while mandatory retirement age is 70. Candidates for judicial positions must pass a public examination and have at least two years of experience as an assistant judge. Professional judges must get a law degree from an accredited university, complete a four-year training programme and serve a two-year court apprenticeship. Together, professional judges and lay assessors deliberate and vote on the appropriate punishment. The public prosecutor is the sole individual with the authority to issue an order on the presentation of charges, to decide on the application of pre-trial custody, and to determine the mechanism for terminating the proceedings. The defendant is deemed innocent unless proven guilty; the prosecution bears the burden of proof. The accused has the right to legal counsel, argumentation, participation in procedural proceedings, appeal of procedural decisions and access to case data. The accused have the right to retain a legal representative (attorney or solicitor) for their defence; if they cannot afford one, they may be eligible for legal aid at public expense.

Typically, court procedures are open to the public and take the form of a recorded oral hearing. After judges and lay assessors discuss and vote on the question of guilt, the legal classification of the act, and other factors, the sentencing decision is made immediately (e.g., civil complaints). This occurs in a single session. There is a separate evidentiary hearing devoted to the topic of the penalty, as well as a separate session in which the penalty is deliberated and decided upon. The verdict is rendered simultaneously on the issue of guilt and punishment, as a whole. The judgement is promptly documented in writing after it is rendered. Additionally, the presiding judge must always verbally announce it in the courtroom. The legislation mandates that the judge and other panel members base sentence decisions on three factors: (1) evidence and its evaluation, (2) scientific principles and (3) personal experience. The accused's confession is regarded an essential piece of evidence, but it is not sufficient to prove guilt. The defendant is not permitted to be questioned under oath. Unlike the Anglo-Saxon procedure, the prosecution and defence do not present evidence separately throughout the trial. Expert testimony is subject to review by the court in accordance with the evidence norms of Polish criminal procedure. The court is therefore not bound by the expert's findings. However, as in most other jurisdictions, the courts accept expert testimony, particularly that of psychiatrists. This is especially true if the mental capacities or possible harm of the perpetrator are considered.

The experiment had a 2 (is the suspect guilty or innocent? guilty vs innocent) × 2 (rating: a single conjunctive rating for both crimes vs. two ratings for each crime) × 3 (participant: judge vs lawyer vs layperson) mixed design (Figure 4.2). The participant variable was a between participants factor, the suspect and rating variables were within participant factors. The conjunction fallacy would be confirmed if there is a main effect of rating. More interestingly, will there be an interaction between rating and participant, with some participant groups less likely to commit the conjunction fallacy? Dual-route approaches to decision-making will lead us to expect that judges would be less likely to commit the fallacy, compared to participants with no legal training.

Each participant was given three criminal case summaries to evaluate. They could select where and when to do the assignment independently. Participants were required to carefully read the descriptions and rate the likelihood that the suspect is guilty on a 10-point scale with the following anchors: 1 = absolutely innocent and 10 = definitely guilty. The subjects required between 20 and 50 minutes to read case descriptions and assign probabilities. For each case, participants either rated the probability of each crime independently (two individual ratings for each crime) or offered a single conjunctive assessment for both crimes. Participants were randomly assigned two versus one scenarios in which they were required to make two separate judgements versus one conjunctive one, or vice versa. Therefore, each participant would need to determine between four and five cases of guilt. For the identical pair of offences, some participants were shown the necessary information in one sequence, while others were shown the information in the opposite order.

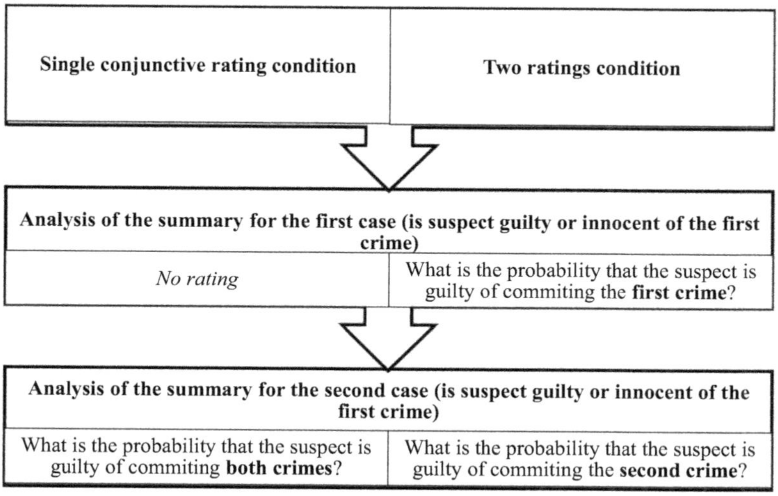

FIGURE 4.2 The design of the study and an outline of the procedure.

Participants

Forty criminal court judges (22 women, 18 men, aged between 29 and 66 years, M = 42.25; SD = 7.66; with professional experience ranging from 3 to 30 years, M=13.5; SD=6.30); 40 prosecutors and attorneys (20 women, 20 men; aged between 28 and 60 years, M = 40.45; SD = 8,35; with professional experience ranging from 1 to 35 years, M=10.05; SD=8.18); and 40 participants without a legal background (21 women, Each individual was recruited in Poland. The first author utilised his professional knowledge and network to discover willing judges, prosecutors and attorneys. The participants who lacked a legal background were also recruited by the first author from among his/her colleagues and friends. The decision to participate or not in this study was entirely voluntary. The subjects were informed that the purpose of the study is to investigate the legal decision-making process and the distinguishing characteristics of judges and lawyers in the evaluation of evidence. Participants got a consent form prior to the study and a debriefing form following the conclusion of the research. Participants were neither compensated nor remunerated for their participation in the study.

Results

We have collected 360 ratings from participants, evenly distributed across the three groups (judges, lawyers, individuals without a legal background). In 170 cases, pairs of crimes were given individual evaluations, whereas 190 pairs of crimes were given a single conjunctive grade. On a 10-point scale, with anchors ranging from 1 (absolutely innocent) to 10 (absolutely guilty), subjects rated the likelihood of a suspect's guilt. When discussing the possibility of guilt in the following section, we assume a linear relationship between these assessments and presumed, subjective probabilities.

We ran a mixed effects model (multilevel linear model in SPSS with random and fixed effects) with probability rating as the dependent variable, profession as a fixed effect between participants independent variables (three levels: judge, lawyer, participants without legal education), rating type as a fixed effect within participants independent variables (three levels: probability for A, probability for B, and probability for the conjunction A&B), and case type (two levels: criminal and civil). The inclusion of two interaction factors, rating × profession and rating × profession × case type, was conceptually driven. It may be thought that case type should be utilised to partition the data file, as ratings for innocent cases and ratings for criminal instances are plainly extremely different. Regarding the key hypothesis of whether a conjunction fallacy occurs or not, a conjunction error involves probability in the same direction for both guilty and innocent situations, that is probability of committing both crimes A and B has to be assessed as higher than committing crime A or crime B. We used a single random participant effect. The random effect was modelled with both intercepts and slopes, but random slopes were only used for the case type fixed effects, in part to avoid overcomplicating the model and in part because the relationship between case type and dependent variable is most likely to be influenced by the participant random effect. Notably, the idea of incorporating a second random effect of specific test item is difficult because crossing these two random effects (participant and specific item) results in numerous cells with a single observation.

The assumption of an unstructured covariance matrix and the estimation of model parameters using maximum likelihood allowed for the use of –2 log likelihood to evaluate nested models. For the sake of brevity, we only consider three models: one without the random effect, one with the random effect but only intercepts, and one with both intercepts and slopes. Each model refinement was very significant; hence the final model was used to evaluate the hypotheses of interest. For the final model, there was a very significant main effect of case type ($F(1,171.5)=403$, $p<.0005$), which is hardly surprising, since it simply suggests that participants behaved differently to situations involving truly guilty suspects as opposed to those involving truly innocent suspects. This primary effect is merely a design validation. In fact, the mean rating for guilty instances was greater than the mean rating for innocent cases (M=8.93, *SE*=.12 vs. M= 3.99, *SE*=.20). There was also a very significant interaction between rating, occupation and case type ($F(8, 207.2)=2.71$, $p<.01$).

By evaluating this triple interaction, we are able to evaluate evidence for a conjunction fallacy, for several professions, and independently for guilty and innocent instances. To do this, we conducted simpler mixed effects models with participant still included as a random effect, but with the random effect's influence now constrained to intercepts; there was only one fixed effect of rating. The *post hoc* approach for estimated means consisted of uncorrected t-tests, as such comparisons were always intended to be tests of a hypothesised conjunction fallacy. For judges, there was an effect of rating when the case type was guilty ($F(2, 64.3)=3.503$, $p=.036$), but no evidence for the conjunction fallacy; rather, probability of committing crime B was rated as higher than both crimes (A&B) (approaching significance, $p=.058$). There was no major effect of rating for lawyers ($F(2,94)=2.33$, p=.10). For participants without a legal background, there was an effect of rating ($F(2,61.9)=5.27$, $p<.01$), and crucially, there was evidence for a double conjunction fallacy, with probability of committing both crimes A and B being higher than probability of crime A and probability of crime B (in both cases, $p<.01$). For when the suspect was truly innocent, there was no influence of rating for judges ($F(2,58.6)=2.08$, $p=.13$), attorneys ($F(2, 72)=0.31$, $p=.74$), and people without a legal education ($F(2,58.4)=1$, $p=.37$). Table 4.1 presents the probability assessments of judges, lawyers and participants without legal background.

In our study of probabilistic evaluation in legal decision-making, a conjunction fallacy was observed. However, only among participants without a legal background, when participants assessed the likelihood that a suspect accused of two crimes was guilty. The conclusion that conjunction fallacy can be detected in legal decision-making is not surprising, given that probabilistic errors have been observed in similar circumstances (Furnham, 1986; Walker et al., 1972). The significant result is the absence of a conjunction fallacy for those with a legal experience. Differences between the groups of participants are also worth noting.

Considering these empirical findings, it is worthwhile to investigate in greater depth to what degree a conjunction fallacy reflects a probabilistic error. Recent attempts to recast the fallacy as a judgement consistent with probability

TABLE 4.1 Conjunction fallacy and guilt probability ratings by different professional groups (on a 1–10 points scale)

Profession	*Guilty–Guilty*						*Innocent–Innocent*					
	Crime A		*Crime B*		*Both Crimes (A&B)*		*Crime A*		*Crime B*		*Both Crimes (A&B)*	
	Mean	*SE*	*Mean*	*SE*	*Mean*	*SE*	*Mean*	*SE*	*Mean*	*SE*	*Mean*	*SE*
Judge	8.66	.25	9.54	.25	8.83	.28	4.25	.48	3.77	.48	3.04	.48
Prosecutor or Attorney	9.06	.22	9.29	.22	8.58	.25	3.42	.50	3.68	.50	3.92	.53
Layperson	8.56⁑	.22	8.56⁑	.22	9.35⁑	0.23	4.68	.43	4.92	.43	4.20	.42

⁑- double conjunction fallacy

theory have been made. We highlight two of these efforts and concentrate on a third. First, proponents of inductive confirmation have argued that in conjunction fallacy experimental paradigms, participants do not mentally compute conjunctions when presented with statements such as bank teller and feminist (in Tversky & Kahneman, 1983, Linda paradigm), but rather with inductive confirmation, the extent to which adding some information (e.g., the Linda story) corroborate or disprove a hypothesis (Tentori et al., 2013). Inductive confirmation is undoubtedly a potent concept in belief updating; nonetheless, there are doubts as to whether a conjunction fallacy paradigm can be recast as an inductive confirmation problem, particularly in the present experiment, where the conjunctive nature of the judgement on both offences was evident Busemeyer et al., 2015). Second, some researchers have recommended an interpretation of normative decision-making based on classical probability theory, but with noise (Costello & Watts, 2014). According to this approach, probabilistic inference includes a process of mental sampling or simulation, which is more prone to error for probabilities with greater complexity, such as conjunctions. However, there is no direct scientific evidence for such a mental sampling process, and the assumption appears especially dubious in instances that are very distinctive and personalised, such as with the crime summaries in the present study. In addition, according to this account, probabilistic fallacies such as the conjunction are still fallacies; the account does not provide a way to reconcile committing a conjunction fallacy with the high expectations for normative decision-making that one would have of lawyers and judges, to say nothing of other professionals.

In general, the quantum probability theory description of the results suggests directions for future research. Quantum probabilities can generate a single conjunction fallacy when the two crime questions are given as incompatible and a double conjunction when the two crime questions are presented as compatible, but there is a "mixing" mental process. There are simple empirical tests, such as order effects, that allow us to experimentally assess if two issues are compatible or incompatible. If they are incompatible, a quantum probability theory model is limited to single conjunctions; if compatible, double conjunction fallacies are allowed, the model must assume that one crime's thoughts and beliefs impact the appraisal of the other. The quantum probability theory modelling demonstrates that the contemplation of one crime influences the consideration of the other crime for (legally) naive individuals. Perhaps for non-expert participants, their assessment of how guilt for one crime affects guilt for another is influenced by their naive familiarity with situations in which a person is either typically guilty or not guilty at all. Notably, despite the descriptive nature of the quantum-like model, it may handle findings that exceed any traditional classical probability theory technique. The current findings give both additional evidence that legal decision-making should be further evaluated for potential fallacies and a preliminary theoretical framework with quantum probability theory that offers possible insights into human behaviour.

It is also worth to comment on differences between the groups of participants. As can be seen in Table 4.1, participants representing different groups of professionals differently assessed probability that suspect is guilty of crime A, of crime B and of both allegations. We have not observed conjunction fallacy for judges and lawyers (prosecutors and attorneys at law), but results show double conjunction error for participants without a legal background. A question may arise: can we explain the differences between the groups of participants in light of the classical and quantum probability theory? According to the classical approach, we should observe similar results for all subjects, either conjunction fallacy for each studied group or lack of fallacy in the experiment. It is not possible to explain observed significant differences without braking basic assumptions of the classical probability theory. Conjunction fallacy among non-professional participants can be explained by a lack of familiarity or a reduction in effort, both of which might increase the likelihood of mismatched representations (Nilsson et al., 2013; Trueblood et al., 2017).

Quantum probability theory can explain observed differences between professionals (judges, prosecutors and attorneys at law) and layperson. The foundations of quantum theory are geometry and linear algebra defined on a Hilbert space (Hilbert spaces are complex vector spaces with particular convergence characteristics). Quantum probability theory begins with what is basically a vector space, such as Figure 4.1s two-dimensional space. A vector can be considered to be a line with a particular direction and length. Such a vector space contains all possible outcomes for questions regarding suspect. For example, there are two possible answers to the question "Is suspect guilty of crime A?" (yes or no), which are represented by two unit-length vectors at a 90° angle to one another, essentially defining the x and y axes. In Figure 4.1, the vertical axis represents the outcome that suspect is guilty of crime A, while the horizontal axis represents the conclusion that the suspect is innocent of crime A. The responses to a different question, such as the guilty question, can be represented by a pair of orthogonal vectors that have been rotated by some angle. In other words, the x and y axes for various queries are related through simple rotations. A vector encoding a question's answer spans a ray, a one-dimensional subspace. Subspaces play a crucial role in quantum theory – it is a method for assigning probabilities to subspaces, and classical probability theory is a method of assigning probabilities to subsets. A person's collection of beliefs regarding these topics is represented by a (unit-length) state vector Ψ (see: Figure 4.1). The probability of a question outcome is calculated by projecting the state vector onto the subspace representing the response and then computing the length squared. Typically, a sequential projection is required to compute the conjunction of two question results, which corresponds to answering each question individually. Consider the case where we are interested in probability of guilty of crime A and guilty of crime B. There is no one ray that corresponds to both being guilty of crime A and being guilty of crime B, as shown in Figure 4.1. Consequently, we must compute probability of guilt of crime A and then guilty of crime B, which requires two steps. First, the state vector is projected onto the "guilty of crime A" ray. Second, this prior projection is projected onto the "guilty of crime B" ray. Probability of finding a suspect guilty of crime B after probability rating of guilt regarding crime A can be represented as the length squared of the final projection. Alternately, assume we wish to calculate probability of guilt of crime B and then probability of guilt of crime A. The state vector is first projected onto the crime B ray, and then the resulting projection is projected onto the crime A ray. Noncommutative sequences of projections are possible. In contrast to the classical probability theory, order is significant, and this noncommutativity results in interference effects. Each question that can be asked is represented in subspaces of the multi-dimensional vector space. Every time a question is answered, the state vector must change its position in a particular way.

Initial position of the vector in the space and alignment of the subspaces assigned to the questions in the multi-dimensional Hilbert space will determine evolution of the belief state and length of the projection on an axis affiliated with a given response to any possible question. Each coordinate of the state vector represents a degree of belief in a statement. There can be different initial positions of the state vector (Ψ) for groups of professionals, different level of initial study participants' conviction that suspect can be guilty of either crime A or B. According to the quantum probability theory assumptions probability of an event can be calculated as squared length of the projection of the state vector on an axis attributed to a question (Busemeyer et al., 2015). In consequence, the closer the belief state is to a certain axis, the higher the likelihood the decision maker assigns to the event represented by that axis (Jaeger & Trueblood, 2019). If the initial location of the state vector is different, not only will the initial probability attributed to an event be different but also the dynamics of the rotation of the state vector in the multi-dimensional Hilbert space will be different. Therefore, length of the projection on the axis affiliated with the last question may different, and probability rating may have a different value.

Differences between the groups of participants can be also explained by distinct characteristics of the vector space assigned to different groups of professionals. In other words, the subspaces for various queries in the multi-dimensional (subspace representing probability of committing crime A and subspace representing probability of committing crime B) can have different alignment. In other words, the subspaces "guilty of crime A" and "guilty of crime B" can have dissimilar properties for each group of professionals, resulting in unique dynamics of the evolution of state vector, and resulting in different length of the projections of the vector on the subspace related to a given question.

Quantum probability theory establishes a formalised notion of how the conjunction fallacy might make logical sense. Thus, the conjunction fallacy is no longer a fallacy and is now the proper response. As a result, we can (almost) conclude that the conjunction fallacy is a reasonable response in certain circumstances (Pothos & Yearsley, 2023).

References

Aerts, D. (2009). Quantum structure in cognition. *Journal of Mathematical Psychology*, Vol. 53, No. 5, pp. 314–348; DOI:10.1016/j.jmp.2009.04.005.

Aerts, D., & Aerts, S. (1995). Applications of quantum statistics in psychological studies of decision processes. *Foundations of Science*, Vol. 1, pp. 85–97.

Aerts, D., & Gabora, L. (2005). A theory of concepts and their combinations ii: A Hilbert space representation. *Kybernetes*, Vol. 34, pp. 192–221; DOI: 10.1108/03684920510575807.

Ariely, D., & Norton, M.I. (2008). How actions create – not just reveal – preferences. *Trends in Cognitive Sciences*, Vol. 12, pp. 13–16; DOI: 10.1016/j.tics.2007.10.008.

Bar-Hillel, M., & Neter, E. (1993). How alike is it versus how likely is it: A disjunction fallacy in probability judgments. *Journal of Personality and Social Psychology, Vol. 65*, No. 6, pp. 1119–1131; DOI: 10.1037/0022-3514.65.6.1119.

Broekaert, J., Basieva, I., Blasiak, P., & Pothos, E.M. (2017). Quantum-like dynamics applied to cognition: A consideration of available options. *Proceedings of the Royal Society A*, Vol. 375, No. 20160387; DOI: 10.1098/rsta.2016.0387.

Bruza, P.D., Wang, Z.J., & Busemeyer, J.R. (2015). Quantum cognition: A new theoretical approach to psychology. *Trends in Cognitive Sciences*, Vol. 19, No. 7, pp. 383–393; DOI: 10.1016/j.tics.2015.05.001.

Busemeyer, J.R., & Bruza, P. (2012). *Quantum Models of Cognition and Decision Making*. Cambridge University Press, Cambridge.

Busemeyer, J.R., Pothos, E.M., Franco, R., & Trueblood, J.S. (2011). A quantum theoretical explanation for probability judgment errors. *Psychological Review*, Vol. 118, No. 2, pp. 193–218; DOI: 10.1037/a0022542.

Busemeyer, J.R., & Wang, Z. (2015). What is quantum cognition, and how is it applied to psychology? *Current Directions in Psychological Science*, Vol. 24, No. 3, pp. 163–169; DOI: 10.1177/0963721414568663.

Busemeyer, J.R., Wang, Z., & Pothos, E. (2015). Quantum models of cognition and decision, [in:] *Oxford Handbook of Computational and Mathematical Psychology*, [eds.] Busemeyer, J.R., Wang, Z., Townsend, J.T., & Eidels, A., Oxford University Press, New York, pp. 369–389.

Busemeyer, J.R., Wang, Z., Pothos, E., & Trueblood, J.S. (2015). The conjunction fallacy, confirmation and quantum theory: Comment on Tentori, Crupi and Russo (2013). *Journal of Experimental Psychology: General*, Vol. 144, No. 1, pp. 236–243; DOI: 10.1037/xge0000035.

Costello, F., & Watts, P. (2014). Surprisingly rational: Probability theory plus noise explains biases in judgment. *Psychological Review*, Vol. 121, pp. 463–480; DOI: 10.1037/a0037010.

Daftary-Kapur, T., Dumas, R., & Penrod, S.D. (2010). Jury decision-making biases and methods to counter them. *Legal and Criminological Psychology*, Vol. 15, pp. 133–154; DOI: 10.1348/135532509X465624.

de Finetti, B., Machi, A., & Smith, A. (1993). *Theory of Probability: A Critical Introductory Treatment*. Wiley, New York.

DeNeys, W., & Goel, V. (2011). Heuristics and biases in the brain: Dual neural pathways for decision making, [in:] *Neuroscience of Decision Making*, [eds.] Vartanian, O., & Mandel, D.R., Psychology Press, New York, pp. 125–141.

Dulany, D.E., & Hilton, D. (1991). Conversational implicature, conscious representation, and the conjunction fallacy. *Social Cognition*, Vol. 9, pp. 85–110; DOI: 10.1521/soco.1991.9.1.85

Elqayam, S., & Evans, J.S.B.T. (2013). Rationality in the new paradigm: Strict versus soft Bayesian approaches. *Thinking and Reasoning*, Vol. 19, pp. 453–470; DOI: 10.1080/13546783.2013.834268.

Furnham, A. (1986). The roubustness of the recency effect: Studies using legal evidence. *The Journal of General Psychology*, Vol. 113, No. 4, pp. 351–357; DOI: 10.1080/00221309.9711045.

Green, E., & Wrightsman, L. (2003). Decision making by Juries and Judges: International perspective, [in:] *Handbook of Psychology in Legal Contexts*, [eds.] Carson, D., & Bull, R., John Wiley and Sons, Chichester, pp. 401–422.

Guthrie, Ch., Rachlinski, J.J., & Wistrich, A.J. (2009). The „Hidden Judiciary": An empirical examination of executive branch justice. *Duke Law Journal*, Vol. 5, pp. 1477–1530.

Haven, E., & Khrennikov, A. (2013). *Quantum Social Science*. Cambridge University Press, Cambridge.

Hogarth, R.M., & Einhorn, H.J. (1992). Order effects in belief updating: The belief-adjustment model. *Cognitive Psychology*, Vol. 24, pp. 1–55; DOI: 10.1016/0010–0285(92)90002-J.

Jaeger, Ch.B., & Trueblood, J.S. (2019). Thinking quantum: A new perspective on decisionmaking in law. *Florida State University Law Review*, Vol. 46, No. 4; pp. 733–806.

Kahneman, D. (2001). *Thinking Fast and Slow*. Penguin, London.

Korf, J. (2015). Quantum and multidimensional explanations in a neurobiological context of mind. *The Neuroscientist*, Vol. 21, No. 4, pp. 345–355; DOI: 10.1177/1073858414542250.

McAuliff, B.D., Nemeth, R.J., Bornstein, B.H., & Penrod, S.D. (2003). Juror decision - making in the twenty - first century: Confronting science and technology in court, [in:] *Handbook of Psychology in Legal Contexts*, [eds.] Carson, D., & Bull, R., John Wiley and Sons, Chichester, pp. 303–372.

Moro, R. (2009). On the nature of the conjunction fallacy. *Synthese*, Vol. 171, pp. 1–24; DOI: 10.1007/s11229.008.9377.8.

Narens, L. (2014). Alternative probability theories for cognitive psychology. *Topics in Cognitive Science*, Vol. 6, pp. 114–120; DOI: 10.1111/tops.12071.

Nilsson, H., Rieskamp, J., & Jenny, M.A. (2013). Exploring the overestimation of conjunctive probabilities. *Frontiers in Psychology*, Vol. 4; DOI: 10.3389/fpsyg.2013.00101.

Oaksford, M., & Chater, N. (2009). Précis of Bayesian rationality: The probabilistic approach to human reasoning. *Behavioral and Brain Sciences*, Vol. 32, pp. 69–120; DOI: 10.1017/S0140525X09000284.

Pothos, E.M., Barque-Duran, A., Yearsley, J., Trueblood, J.S., Busemeyer, J.R., & Hampton, J.A. (2015). Progress and current challenges with the quantum similarity model. *Frontiers in Psychology*, Vol. 6; DOI: 10.3389/fp-syg.2015.00205.

Pothos, E.M., & Busemeyer, J.R. (2009). A quantum probability explanation for violations of "rational" decision theory. *Proceedings of the Royal Society B: Biological Sciences*, Vol. 276, No. 1665, pp. 2171–2178; DOI: 10.1098/rspb.2009.0121.

Pothos, E.M., & Busemeyer, J.R. (2013). Can quantum probability provide a new direction for cognitive modeling? *Behavioral and Brain Sciences*, Vol. 36, pp. 255–327; DOI: 10.1017/S0140525X12001525.

Pothos, E.M., Busemeyer, J.R., Shiffrin, R.M., & Yearsley, J.M. (2017). The rational status of quantum cognition. *Journal of Experimental Psychology: General*, Vol. 146, No. 7, pp. 968–987; DOI: 10.1037/xge0000312.

Pothos, E.M., & Busemeyer, J.R. (2022). Quantum Cognition. *Annual Reviews in Psychology*, Vol. 73, pp. 749–778; DOI: 10.1146/annurev-psych-033020-123501.

Pothos, E.M., Shiffrin, R.M., & Busemeyer, J.R. (2013). The dynamics of decision making when probabilities are vaguely specified. *Journal of Mathematical Psychology*, Vol. 59, pp. 6–17; DOI: 10.1016/j.jmp.2013.09.001.

Pothos, E.M., & Trueblood, J.S. (2015). Structured representations in a quantum probability model of similarity. *Journal of Mathematical Psychology*, Vol. 64–65, pp. 35–43; DOI: 10.1016/j.jmp.2014.12.001.

Pothos, E.M., & Yearsley, J.M. (2023). Quantum cognition: Quo vadis? [in:] *The Quantum – Like Revolution,* [eds.] Plotnitsky, A., & Haven, E., Springer Nature, Cham, pp. 281–297; DOI:10.1007/978-3-031-12986-5_14.

Saks, M.J., & Thompson, W.C. (2003). Assessing evidence: Proving facts, [in:] *Handbook of Psychology in Legal Contexts*, [eds.] Carson, D., & Bull, R., John Wiley and Sons, Chichester, pp. 329–345.

Sides, A., Osherson, D., Bonini, N., & Viale, R. (2002). On the reality of the conjunction fallacy. *Memory & Cognition*, Vol. 30, pp. 191–198; DOI: 10.3758/BF03195280.

Simon, D, (2012). *In doubt. The Psychology of the Criminal Justice Process.* Harvard University Press, Cambridge, MA, London.

Sloman, S.A. (1996). The empirical case for two systems of reasoning. *Psychological Bulletin*, Vol. 119, pp. 3–22.

Spellman, B.A., & Tenney, E.R. (2010). Credible testimony in and out of court. *Psychonomic Bulletin & Review*, Vol. 17, No. 2, pp. 168–173; DOI: 10.3758/PBR.17.2.168.

Tenenbaum, J.B., Kemp, C., Griffiths, T.L., & Goodman, N. (2011). How to grow a mind: Statistics, structure, and abstraction. *Science*, Vol. 331, pp. 1279–1285; DOI: 10.1126/science.1192788.

Tenney, E.R., Small, J.E., Kondrad, E.L., Jaswal, V.K., & Spellman, B.A. (2011). Accuracy, confidence, and callibration: How young children and adults assess credibility. *Developmental Psychology*, Vol. 47, No. 4, pp. 1065–1077; DOI: 10.1037/a0023273.

Tentori, K., & Crupi, V. (2012). On the conjunction fallacy and the meaning of and, yet again: A reply to Hertwig, Benz, and Krauss (2008). *Cognition*, Vol. 122, pp. 123–134; DOI: 10.1016/j.cognition.2011.09.002.

Tentori, K., Crupi, V., & Russo, S. (2013). On the determinants of the conjunction fallacy: Probability versus inductive confirmation. *Journal of Experimental Psychology: General*, Vol. 142, No. 1, pp. 235–255; DOI: 10.1037/a0028770.

Trueblood, J.S., & Busemeyer, J.R. (2010). A quantum probability account for order effects on inference. *Cognitive Science*, Vol. 35, pp. 1518–1552.

Trueblood, J.S., & Busemeyer, J.R. (2011). A quantum probability model of casual reasoning. *Frontiers in Cognitive Science*, Vol. 3; DOI: 10.3389/fp-syg.2012.00138.
Trueblood, J.S., Yearsley, J.M., & Pothos, E.M. (2017). A quantum probability framework for human probabilistic inference. *Journal of Experimental Psychology: General*, Vol. 146, No. 9, pp. 1307–1341; DOI: 10.1037/xge0000326.
Tversky, R., & Kahneman, D. (1983). Extensional versus intuitive reasoning: The conjunctive fallacy in probability judgment. *Psychological Review*, Vol. 90, pp. 293–315.
Tversky, R., & Kahneman, D. (1986). Rational choice and the framing of decisions. *The Journal of Business*, Vol. 59, No. 4, Part 2: The Behavioral Foundations of Economic Theory, pp. 251–278.
Walker, L., Thibaut, J., & Andreoli, V. (1972). Order of presentation at trial. *Yale Law Journal*, Vol. 82, pp. 216–226.
Wang, Z., Solloway, T., Shiffrin, R.M., & Busemeyer, J.R. (2014). Context effects produced by questions orders reveal quantum nature of human judgment. *PNAS*, Vol. 111, No. 26, pp. 9431–9436; DOI: 10.1073/pnas.1407756111.
White, L.C., Barque – Duran, A., & Pothos, E.M. (2016). An investigation of a quantum probability model for the constructive effect of affective evaluation. *Philosophical Transactions A*, Vol. 374, No. 20150142; DOI: 10.1098/rsta.2015.0142.
White, L.C., Pothos, E.M., & Busemeyer, J.R. (2014). Sometimes it does hurt to ask: The constructive role of articulating impression. *Cognition*, Vol. 133, pp. 48–64; DOI: 10.1016/j.cognition.2014.05.015.
Yukalov, V.I., & Sornette, D. (2014). Conditions for quantum interference in cognitive sciences. *Topics in Cognitive Science*, Vol. 6, pp. 79–90; DOI: 10.1111/tops.12065.

5

CRIMINAL PROCEDURE AND QUANTUM ZENO EFFECT

The quantum Zeno effect

Zeno of Elea, an ancient Greek philosopher, investigated apparent paradoxes through the use of mental experiments. For almost 2,500 years, Zeno's paradoxes of motion have attracted the interest of philosophers, mathematicians and scientists (Alper & Bridger, 1997). In one of the experiments, Zeno pictured an arrow soaring across the air. He reasoned that the arrow remains stationary at each moment. How can an arrow traverse through space if its trajectory is entirely made of motionless moments? There cannot be motion. The paradox can be simply pictured with the use of another simple analogy: if one atom is not visible, then two atoms are not visible, and then three atoms are not visible, does this imply that the visible world does not exist? (Belinsky, 2018). For centuries philosophers, mathematicians and scientists were searching for answer to the question. Finally, calculus made evident the solution to the original Zeno's arrow problem. If we study the arrow's motion over some interval of time, we find that it has a finite speed. This time period can be made infinitely small, as close to 0 as we like, and the arrow's speed will remain limited. As a result, we can compute its finite speed at each given moment and demonstrate that the arrow does really move through the air.

The ancient arrow paradox of Zeno has subsequently evolved into a quantum thought experiment, the "quantum Zeno effect," by which we can freeze the state of quantum systems by watching them continually. George Sudarshan and Baidyanath Misra of the University of Texas in 1977 published a paper entitled "*The Zeno's paradox in quantum theory.*" The authors of the study considered a radioactive particle that is unstable and has a chance of decaying. Using quantum theory, scientists determined that if the particle is

DOI: 10.4324/9781003402480-5

constantly watched, it should never decay. Measurement of a quantum system hinders its transition to another state (Belinsky, 2018). This phenomenon is also called the watchdog effect (Kraus, 1981). The quantum Zeno paradox has been experimentally confirmed and explained by quantum theory. Freezing a quantum state by continuous observations is a consequence of the quantum theory collapse postulate (von Neumann's postulate), which entails that measurements reset the state to be consistent with the measured outcome (Harrington et al., 2017; von Neumann, 1955; Wheeler & Zurek, 1983). Why does a quantum arrow halt when its localized state with a momentum greater than zero is regularly examined? The solution is inextricably linked to the Heisenberg uncertainty principle, which asserts that a quantum system with a definite momentum cannot be precisely localized: its location is uncertain up to the breadth of its quantum wave. When our arrow travels a distance less than this width, its quantum state becomes a superposition of the initial quantum state and a new state. A measurement determining whether or not the arrow is still in its original state may or may not detect it there. The likelihood that the arrow has moved since the last measurement is related to the square of the time that has transpired since the last measurement, and it decreases dramatically if measurements are performed often. Increasing the number of measurements over a certain amount of time eventually decreases the probability to zero – the quantum Zeno effect takes effect (Vaidman, 2008). Frequent measurements on the system result in projecting the system back to its initial state. In other words, the wave function of the system is repeatedly collapsed, and the evolution is frozen (Ghose, 1999). It is also necessary that the time interval between successive measurements must be much shorter than the critical time of coherent evolution of the system, called the Zeno time. For decays of the particles, Zeno time is the time of coherent evolution before the irreversible exponential decay sets in. For most decays, this is extremely short and hard to detect (Ghose, 1999).

It is worth to note that there is a significant distinction between continually measuring a quantum state and the classic Zeno's arrow thought experiment. Seeing quantum systems has a significant impact on the system's state, whereas observing an arrow flying through the air has no effect. Also, it is worth to mention the anti-Zeno effect which occurs when repeated measurements speed up a decaying process. Anti-Zeno effect was originally detected (together with the Zeno effect) in a tunnelling experiment with a cold atomic vapour (Fischer et al., 2001; Harrington et al., 2017).

As the Zeno effect is a general consequence of the laws of quantum mechanics, one would expect to see it in a variety of physical experiments and observations. Indeed, this is the case. Several examples have been observed in physical experiments, such as the existence of cloud chamber tracks and the inhibition of atomic transitions, the stability of a sugar solution due to numerous collisions with solvent molecules, and the stability of a neutron within a nucleus (Ghose, 1999). Quantum Zeno paradox has also influenced research on causality (Belinsky, 2018), reversible and irreversible quantum evolutions (Gustafson, 2005), continuous observation (Roy, 2001), quantum chaos (Saito, 2004), superconducting qubits and structured noise baths (Harrington et al., 2017), and quantum dynamic games (Kolokoltsov, 2022).

Atmanspacher et al. (2004, 2010) noticed that modelling the dynamics of the bistable perception of ambiguous visual stimuli such as the Necker cube requires a quantum theoretical framework that is not limited to the validity domain of conventional quantum physics. The key concept proposed by the researchers was to analyse the perception process as the evolution of an unstable two-state system, which produced a Necker-Zeno effect, analogous to the quantum Zeno phenomenon. Theoretically, a quantitative relationship is deduced between the relevant time scales. Cognitive time scales relevant for bistable perception are found to objectively satisfy this relationship (Atmanspacher et al., 2004, 2010).

The role of reasoning in legal practice has been addressed in many articles and books. Researchers usually focus on logic, uncertainty, biases, probability or search for ways to model legal arguments and their assessment (i.e. Aitken et al., 1996; Dror & Charlton, 2006; Fenton et al., 2013; Fenton & Neil, 2011; Jaeger & Trueblood, 2019; Oaksford & Chater, 2007, 2010). The most influential approach has been traditional, Bayesian probability theory. Bayesian principles are supported by powerful reasoning (such as the Dutch book theorem) and robust, ingrained intuition. Given the relevant data, Bayesian models are regarded normative, as they represent how decisions should be made. Bayesian principles are frequently motivated by adaptive considerations that are equally relevant to human and non-human decision-makers, despite the fact that research on rationality traditionally focuses on human decision-making (Yearsley & Pothos, 2016). Bayesian cognitive models have typically proved successful (Tenenbaum et al., 2011). Researchers have on occasion noticed a persistent gap between Bayesian prediction and behaviour. The influential Tversky–Kahneman research tradition is most closely associated with these discoveries (Tversky & Kahneman, 1983). These discoveries have produced profound theoretical divisions, with some academics denying formal probability theory's significance in cognitive modelling outright. Some scholars have investigated if quantum theory can serve as an alternative formal foundation for cognitive theory (i.e. Aerts & Aerts, 1995; Busemeyer & Bruza, 2012; Haven & Khrennikov, 2013; Yearsley & Pothos, 2016).

Commonly used models in psychology assume that the cognitive system changes from moment to moment, but at any given moment, it is in a fixed state with respect to a decision to be made. Suppose, for example, that a judge has just heard conflicting evidence from the prosecutor and the defence and must choose between two mutually exclusive and exhaustive hypotheses: suspect is guilty or suspect is innocent. A Bayesian model would provide a probability distribution over the two hypotheses: a probability of guilt, taking into account collected evidence and a probability of innocence, given proved facts. Hence, the judge's subjective likelihood with regard to the question of guilt or innocence is represented by a point between zero and one on the probability scale at each given time. This likelihood may fluctuate from moment to moment, resulting in a distinct progression of guilt probability across time. Yet, at any given time, this subjective probability either favours guilt (is higher for guilt), innocence (is higher for conclusion that suspect is innocent) or is exactly equal to 50%. A judge cannot support both guilt and innocence at the same time (Busemeyer et al., 2015). Quantum theory, on the other hand, assumes that judges are always in an indeterminate (superposition) condition during deliberations. Being in an indeterminate position, the decision-maker does not certainly favour guilty nor does the judge definitely favour innocence. Instead, the judge is in a superposition state that leaves him or her conflicted, ambiguous, perplexed, or doubtful regarding the status of guilt. At a given time, the probability of stating guilt may be greater than the probability of stating innocence, and these probabilities may fluctuate from one minute to the next; yet, either hypothesis may be selected at any given moment. In quantum theory, there is no one trajectory or path sampled across time before a decision is made. A judge would be required to commit to either guilt or innocence, when asked to render a verdict (Busemeyer et al., 2015).

In classical probability theory, Bayes' rule states that if a person assigns a probability of zero or one to any hypothesis, then no additional evidence can change the probability from these extremes. This rule appears to conflict with human reasoning. Basieva et al. (2017) conducted two experiments in which participants judged the likelihood of guilt for group of possible criminals. Early evidence prompted participants to give zero (or a very low) probability to one candidate, but later evidence moved these initial zero probabilities to a high degree of guilt. Violation of the classical Bayesian updating of belief was observed (Basieva et al., 2017). When strong evidence was added to an update, the level of confidence could go from zero, which is almost like denying the possibility, to almost 100%. This so-called zero priors paradox is an intriguing example of the collapse principle in action. In the quantum model proposed by Basieva et al. (2017), it is possible to move from zero priors to a nonzero probability of guilt by utilizing the manner in which a prior decision modifies the mental state.

Yearsley and Pothos (2016) also explored the quantum Zeno paradox in cognition, considering the prediction of the collapse principle. In their experiment, participants were asked to determine the chance that a defendant was guilty of a crime after receiving a series of evidence; the defendant was initially considered to be innocent. In various conditions, participants were asked to evaluate certain number of items of information, with that number varying among conditions. According to the collapse principle, the greater the number of intermediate judgements (smaller amount of information at each stage), the greater the likelihood that the judgements will keep collapsing the state back to the initial one, resetting the person's perception of that cases. This is the quantum Zeno effect, which in a decision-making context refers to the notion that the frequency of intermediate assessments reduces the rate of opinion change (colloquially, a watched pot never boils). Results of the study by Yearsley and Pothos (2016) confirmed that the quantum Zeno effect can be observed in decision-making.

Study on the criminal trial dynamics

According to the provisions of Polish criminal procedure (and similarly to rules of conduct in criminal cases of the other countries) a court may use detention against a suspect if it is required to ensure the proper conduct of the proceedings. Occasionally, detention may also be used to prevent the defendant from committing a serious new crime. The public prosecutor may submit a request for provisional detention or custody. In the application, evidence indicating a high probability that the accused has committed the offence and circumstances indicating a threat to the integrity of the proceedings must be detailed. Detention is only permissible if the gathered evidence indicates a high likelihood that the accused has committed an offence. During pending proceedings, the court may choose to revoke or prolong the custody. The decision regarding application or extension may be based on the conclusions drawn from the evidence disclosed to the defendant and defence counsel.

If a request for provisional detention is made when the substantive case is already pending, the petition is filed with the court reviewing the case at that moment. An exception is the Supreme Court, which does not have jurisdiction over the subject of security for claims – in proceedings before the Supreme Court matters of security are decided by the court competent in the first instance. The court's decision regarding the application of the preventive measure shall be in the form of an order. The court determines the length of the detention and justifies its decision with evidence of the suspect's commission of an offence, the demonstration of circumstances indicating the existence of threats to the proper conduct of the proceedings or the possibility of the accused committing a new, grave offence if the precautionary measure is not applied, as well as the specific grounds for its application and the necessity of the measure. In the case of the detention, it must also be clarified why an alternative preventive measure was not deemed adequate.

A decision regarding a preventive measure is appealable. The preventive measure must be immediately revoked or modified if the reasons for its imposition no longer exist or if new reasons justifying its revocation or modification emerge. The accused may request the revocation or modification of the preventive measure at any time. At this point, it is worth to note that the application of arrest is not equivalent to establishing the accused's guilt. In the case of an arrest, the court determines that there is a high likelihood that the alleged act was committed, whereas guilt is established beyond a reasonable doubt in the case of a conviction.

Throughout the proceedings, the court may be required to repeatedly evaluate the gathered evidence and the reasons for further detention, as well as provide a preliminary assessment of the suspect's likelihood of committing the charged acts. The court may be required to make multiple decisions concerning the need for further detention during the criminal proceedings. And new facts and evidence that emerge during the course of the trial can cause the judge to change a previous decision. In our opinion, court rulings regarding provisional measures are an interesting example of a dynamic and complex decision-making process, with multiple possible measurements of the judges' cognitive state and opinions. An analysis of the quantum theory literature raises the question of whether the quantum Zenon effect can be observed in the legal settings when numerous intermediate decisions related to the course of the criminal trial (regarding interim measures) are being made.

In classical probability theory it is assumed that judgements arise from definite states. A person who makes decisions always possesses a mental picture of the relevant probability. This opinion may alter frequently as new information becomes available, but the decision-maker always has distinct opinions. When a measurement occurs (i.e., when a court must determine on interim measures), the probabilistic judgements that are already in the mind of the judges are simply read out and used to assist the decision-making. In the quantum view, it is not only assumed that judgements arise from indefinite states (decision-makers can remain in a state of "I don't know"), but also predicts that changes in evidence based opinions (such as whether a criminal defendant is guilty or innocent) will be slowed down by repeated judgements (Bruza et al., 2015; Bruza & Busemeyer, 2012; Busemeyer & Wang, 2015; Yearsley & Pothos, 2016). Quantum principles suggest that when someone's opinion on a topic is "measured," articulated or consciously identified – it becomes less likely to change, even in the presence of accumulating evidence to the contrary (Jaeger & Trueblood, 2019; Yearsley & Pothos, 2016).

Quantum decision-making proposes that decision-makers evaluate probabilities based on quantum probability theory (Bruza et al., 2015). The quantum Zeno effect is a consequence of the collapse postulate of quantum theory, which states that measurements (or in decision-making, judgements) reset the state to be consistent with the measured outcome. If there is quantum structure in cognition, then the collapse hypothesis must also be supported by experiment (Yearsley & Pothos, 2016). So, the goal of our work was to investigate a quantum Zeno effect in human decision-making and to give evidence supporting the use of quantum principles in decision theory, at least in some circumstances. A noteworthy a priori prediction is that repeated, continual judgements will reduce the rate of opinion change among decision-makers.

Yearsley and Pothos (2016) developed a model designed to illustrate an effect for opinion change in an experiment focused on assessment of suspect's guilt. The proposed model can be by analogy applied to study decisions made by judges ruling on provisional detention or custody. The main differences between the previous study by Yearsley and Pothos (2016) and our experiment, is that naive participants rated probability of guilt, whereas we have conducted study on the group of criminal courts judges deciding on the need for provisional measures.

Yearsley and Pothos (2016) proposed to model the opinion change by considering a two dimensional quantum system, whose state space is spanned by two orthogonal states I (innocent) and G (guilty), corresponding to the beliefs regarding suspect's guilt. Presentation of evidence can be represented by a rotation of the state, such that an initial state of the decision-maker evolves in time, following presentation of evidence. Following assumptions of the quantum theory it can be expected, that as the number of measurements increases, there is a decreasing probability that the system will change from I to G. The probability that the system will change state vanishes, after presentation of the large number of pieces of evidence. Even though the derivation leading to the equation proposed by Yearsley and Pothos (please see Yearsley & Pothos, 2016) allows to predict a weakened quantum Zeno effect, as a slowing down (in a specific way) of the evolution of the measured opinion state. The authors of the model point out two assumptions that need to be relaxed – reliability of the measurement (there is a small probability that a participant will incorrectly provide a response not matching her/his cognitive state) and time independent unitary evolution (since the evolution of cognitive variables is of interest, weight given by participants to pieces of evidence may depend on their position in the sequence, only approximately unitary evolutions can be assumed). Yearsley and Pothos (2016) offer an equation allowing to determine probability that the cognitive state of the participant will be consistent with previous change as a result of presentation of the additional evidence. The equation is based on few terms: the first term is the probability that the state never changes, multiplied by the probability that the N imperfect measurements all come out in the expected way (i.e., that participant will express opinion that suspect is innocent); the second term represents the probability that the state changes between the second to last and last measurements, but that the last measurement fails to detect this change; further terms either represent earlier changes in the state, and so more failed detections, or the state changing back to innocent from guilty (the probability for this last possibility is expected to be negligible for other reasons, since a participant who thinks suspect is guilty is very unlikely to revert and respond that suspect is innocent, after seeing more guilty evidence) (for more details please see supplementary material to Yearsley & Pothos, 2016; as well as, Yearsley & Busemeyer, 2016).

Material and methods

The research was conducted in Poland using Polish criminal courts' records. We have been granted access to 20 real-life criminal cases, in which provisional detention has been ruled. We have selected four cases with the largest number of rulings on the use of detention throughout the proceedings. Cases came from different court district than the participants of the study to minimalize risk of case familiarity. Two independent raters (criminal judges) assessed evidence and prepared proposals for case summaries – selected 36 most significant proofs and facts (18 incriminating and 18 exonerating) for each case. Of these, 12 pieces of evidence and findings (six indicating guilt and six indicating innocence) on which significance the raters agreed the most (had the highest concordance), were chosen. On this basis case summaries were prepared. Content was split in six blocks, each included description of one new evidence or finding indicating suspect's guilt and one piece indicating innocence, this allowed to balance the arguments for and against the use of detention and focus the study on the decision-making dynamics (and not the legal assessment). The significance of the various pieces of evidence or information forming one block was also counter balanced, so for each part incriminating equally important exonerating circumstances were presented. As a result, case summaries were optimized, the impact of factual circumstances of an individual case on participants' decisions was minimalized and the main effect was enhanced.

The participants were first provided with some basic information about the study and a consent form, complying with the ethics committee's guidelines. The participants were informed that the study is focused on legal decision-making and dynamics of the criminal procedure and its aim is to investigate influence of evidence and findings on use of provisional measures. Each of the participants was given summaries of the four cases prepared for the experiment. Different cases were presented in different order for different participants, but the order of the blocks was fixed. Each of the participants rated all four cases, but presentation and type of rating varied between the cases and between the subjects. We have collected overall 180 ratings. Case summaries were divided in six blocks (with one incriminating evidence or fact and one exonerating evidence or fact). For each of the four cases, 30 participants rated the necessity to order detention after reading all six blocks (12 pieces of information), and there were 30 ratings for each of the cases, requested at intervals of either 2, 4, 6, 8, 10 or 12 pieces of evidence. Participants had to make decision once (after reading all six blocks and 12 pieces of information), twice, three, four, five or six times (after each block of evidence). Each participant rated different cases in variations, i.e. first case once, second case four times, third case five times and fourth case six times.

Participants

The study was conducted on a group of 45 criminal division judges, 16 women and 29 men, aged between 38 and 62 years (M = 48.13; SD = 8.35) with professional experience ranging from 6 to 33 years (M=17.92; SD=11.26). Participation in this research was optional. For the study author recruited judges previously participating in trainings and workshops on content analysis and veracity assessment. Participants were informed that the purpose of the study was to investigate legal decision-making in the provisional detention cases. Prior to the study, participants received a consent form and, following its conclusion, a debriefing form. The subjects were not compensated for their participation.

Results

The main purpose of the study was to examine whether the quantum Zeno effect can be observed when judges of the criminal division make decision on provisional detention. Following predictions of the quantum probability theory we expect steady decline of the probability of judge's ruling on detention progressing with each successive assessment. We ran two a within- and between-subjects design repeated measures ANOVA. In the first model: with number of decisions made by the judge as the within-subject variables, participant as the between-subject factor and probability of decision's change as the dependent variable. The analysis showed insignificant effect for the two-way interaction of participant and number of decisions made by the participant ($F(5,45)=1.34$, $p=.12$). In the second model: with number of decisions made by the judge as the within-subject variables, case as the between-subject factor and probability of decision's change as the dependent variable. There was no effect for the two-way interaction of case and number of decisions made by the participant ($F(3,60)=1.22$, $p=.20$).

We have found simple main effect of number of decisions made by the judge ($F(5,45)=3.87$, $p<.05$). Comparison of the mean probabilities of decision's change for successively increasing number of rulings show that intermediate assessments reduce the rate of opinion change ($M_{FirstRating}=.41$, $SD_{FirstRating}=.17$; $M_{secondRating}=.36$, $SD_{SecondRating}=.19$; $M_{ThirdRating}=.31$, $SD_{ThirdRating}=.19$; $M_{FourthRating}=.28$, $SD_{FourthRating}=.19$; $M_{FifthRating}=.26$, $SD_{FifthRating}=.20$; $M_{SixthRating}=.26$, $SD_{SixthRating}=.20$). Changes in dynamics of the judges' decision-making process are presented in Figure 5.1.

Detailed comparisons of differences between the mean probabilities of change of participants' decisions to place suspect in custody, depending on which decision in turn the judge made is presented in Table 5.1.

It is crucial to study opinion changes, especially in the legal settings and search for ways to deepen our understanding of the dynamics of opinion evolution.

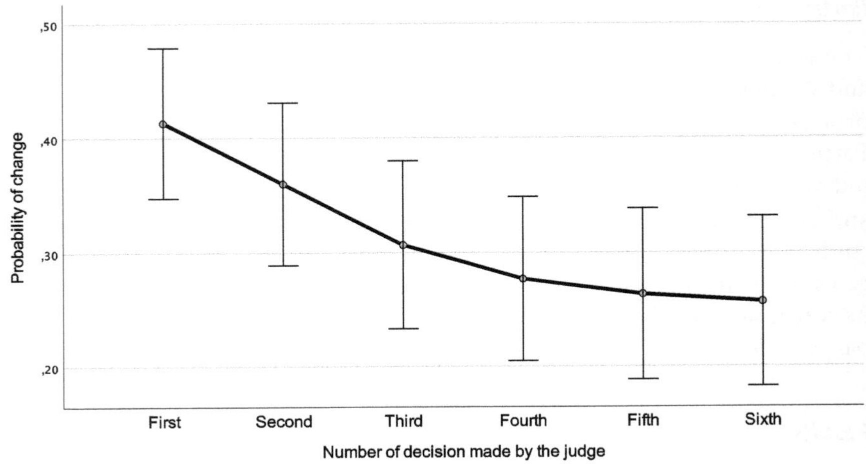

FIGURE 5.1 Mean probability of decision's change for different stages of decisions.

TABLE 5.1 Comparison of mean probabilities of change for different number of pieces of evidence after which judgement occurs

Number of rulings		*First decision*	*Second decision*	*Third decision*	*Fourth decision*	*Fifth decision*
		$M_i - M_j$	$M_i - M_j$	$M_i - M_j$	$M_i - M_j$	$M_i - M_j$
Second decision	$M_i - M_j$	.05				
Third decision	$M_i - M_j$	.11*	.05			
Fourth decision	$M_i - M_j$	.14*	.08*	.03		
Fifth decision	$M_i - M_j$	.15*	.10*	.04	.01	
Sixth decision	$M_i - M_j$	.16**	.11*	.05	.02	.01

*pairwise comparison indicating significant differences with $p<.05$; **pairwise comparison indicating significant differences with $p<.001$

Collected data suggest that not only that arguments, evidence and findings determine court rulings, but also that judge's assessment can be affected by frequent judgements. Results confirm striking prediction following from a structural feature of the quantum theory, the state vector's collapse postulate. Evolution of the system can be significantly slowed or even frozen as a result of repeated measurement. To put it another way, when a person's opinion is it becomes less likely to change, even in the face of mounting evidence to the contrary. Each time a judgement is made, the person's views are "reset" to reflect it. Current study confirms previous findings of Yearsley and Pothos (2016), as well as Basieva et al.'s (2017) and predictions of Jaeger and Trueblood (2019).

The quantum Zeno effect observed in this research has structural and tactical legal ramifications. One can picture, for instance, a criminal defence attorney asking jurors to evaluate each piece of the prosecution's evidence as it is presented to determine whether it convinces them that the defendant is guilty. One may also expect that jurors with specific characteristics, such as a propensity for introspection and belief reevaluation, may be more susceptible to the quantum Zeno effect, and that inquiries pertaining to such characteristics may be relevant during *voir dire* (Jaeger & Trueblood, 2019). It can be also expected that the judge will consistently impose detention, even if evidence and findings collected during the proceedings would favour reversing the prior order. The quantum Zeno effect is not on the conventional list of heuristics and biases described in the literature, even though it is both very "irrational" from a classical perspective and highly relevant to law. This is probably because its existence is predicted by the quantum decision-making, and not by the classical probability theory.

It is also worth noting that despite the counterbalance of the various pieces of evidence and information constituting a single block (equal significance of incriminating and exonerating circumstances), the likelihood that the judge will reverse his or her previous decision was significantly lower than chance.

References

Aerts, D., & Aerts, S. (1995). Applications of quantum statistics in psychological studies of decision processes. *Foundations of Science*, Vol. 1, pp. 85–97.

Aitken, C.G.G., Connolly, T., Gammerman, A., Zhang, G., Bailey, D., Gordon, R., & Oldfield, R. (1996). Statistical modelling in specific case analysis. *Science & Justice*, Vol. 36, No. 4, pp. 245–255; DOI: 10.1016/S1355-0306(96)72610-2

Alper, J.S., & Bridger, M. (1997). Mathematics, models and Zeno's paradoxes. *Synthese*, Vol. 110, pp. 143–166.

Atmanspacher, H., Filk, T., & Roemer, H. (2004). Quantum Zeno features of bistable perception. *Biological Cybernetics*, Vol. 90, pp. 33–40; DOI: 10.1007/s00422-003-0436-4.

Atmanspacher, H., & Filk, T (2010). A proposed test of temporal nonlocality in bistable perception. *Journal of Mathematical Psychology*, Vol. 54, No. 3, pp. 314–321; DOI: 10.1016/j.jmp.2009.12.001

Basieva, I., Pothos, E.M., Trueblood, J., Khrennikov, A., & Busemeyer, J.R. (2017). Quantum probability updating from zero prior (by-passing Cromwell's rule). *Journal of Mathematical Psychology*, Vol. 77, pp. 58–69; DOI: 10.1016/j.jmp.2016.08.005.

Belinsky, A.V. (2018). On the violation of causality in quantum experiments. *Moscow University Physics Bulletin*, Vol. 73, No. 3, pp. 252–262; DOI: 10.3103/S0027134918030037.

Bruza, P.D., & Busemeyer, J.R. (2012). *Quantum Models of Cognition and Decision*. Cambridge University Press, Cambridge.

Bruza, P.D., Wang, Z.J., & Busemeyer, J.R. (2015). Quantum cognition: A new theoretical approach to psychology. *Trends in Cognitive Sciences*, Vol. 19, No. 7, pp. 383–393; DOI: 10.1016/j.tics.2015.05.001.

Busemeyer, J.R., & Bruza, P.D. (2012). *Quantum Models of Cognition and Decision.* Cambridge University Press, Cambridge.
Busemeyer, J.R., & Wang, Z. (2015). What is quantum cognition, and how is it applied to psychology? *Current Directions in Psychological Science*, Vol. 24, No. 3, pp. 163–169; DOI: 10.1177/0963721414568663.
Busemeyer, J.R., Wang, Z., & Pothos, E. (2015). Quantum models of cognition and decision, [in:] *Oxford Handbook of Computational and Mathematical Psychology*, [eds.] Busemeyer, J.R., Wang, Z., Townsend, J.T., & Eidels, A., Oxford University Press, New York, pp. 369–389.
Dror, I.E., & Charlton, D. (2006). Why experts make errors. *Journal of Forensic Identification*, Vol. 56, No. 4, pp. 600–616.
Fenton, N., & Neil, M. (2011). Avoiding legal fallacies in practice using Bayesian networks. *Australian Journal of Legal Philosophy*, Vol. 36, pp. 114–150.
Fenton, N., Neil, M., & Lagnado, D.A. (2013). A general structure for legal arguments about evidence using Bayesian networks. *Cognitive Science*, Vol. 37, pp. 61–102; DOI: 10.1111/cogs.12004.
Fischer, M.C., Gutierrez-Medina, B., & Raizen, M.G. (2001). Observation of the quantum Zeno and Anti-Zeno effects in an unstable system. *Physical Review Letters*, Vol. 87; DOI: 10.1103/PhysRevLett.87.040402
Ghose, P. (1999). *Testing Quantum Mechanics on New Ground.* Cambridge University Press, Cambridge.
Gustafson, K. (2005). Bell and Zeno. *International Journal of Theoretical Physics*, Vol. 44, No. 11, pp. 1933–1942; DOI:10.1007/s10773-005-8971-5.
Harrington, P.M., Monroe, J.T., & Murch, K.W. (2017). Quantum zeno effects from measurement controlled qubit-bath interactions. *Physical Review Letters*, Vol. 118, No. 240401; DOI: 10.1103/PhsyRevLett.118.240401.
Haven, E., & Khrennikov, A. (2013). *Quantum Social Science.* Cambridge University Press, Cambridge.
Jaeger, Ch.B., & Trueblood, J.S. (2019). Thinking quantum: A new perspective on decisionmaking in law. *Florida State University Law Review*, Vol. 46, No. 4, pp. 733–806.
Kolokoltsov, V.N. (2022). Dynamic quantum games. *Dynamic Games and Applications*, Vol. 12, pp. 552–573; DOI: 10.107/s13235-021-00389-w.
Kraus, K. (1981). Measuring processes in quantum mechanics I. Continuous observation and the watchdog effect. *Foundations of Physics*, Vol. 11, Nos. 7/8, pp. 547–576.
Misra, B., & Sudarshan, E.C.G. (1977). The Zeno's paradox in quantum theory. *Journal of Mathematical Physics*, Vol. 18, pp. 756–763.
Oaksford, M., & Chater, N. (2007). *Bayesian Rationality: The Probabilistic Approach to Human Reasoning.* Oxford University Press, Oxford; DOI: 10.1093/acprof:oso/9780198524496.001.0001
Oaksford, M., & Chater, N. (eds.) (2010). *Cognition and Conditionals: Probability and Logic in Human Thinking.* Oxford University Press, Oxford; DOI: 10.1093/acprof:oso/9780199233298.001.0001
Roy, S.M. (2001). Quantum Zeno and anit-Zeno paradoxes. *Pramana – Journal of Physics*, Vol. 56, Nos. 2/3, pp. 169–178.
Saito, N. (2004). Theory of quantum measurement in terms of quantum chaos. *Journal of the Physical Society of Japan*, Vol. 73, No. 6, pp. 1476–1481; DOI: 10.1143/JPSJ.73.1476.

Tenenbaum, J.B., Kemp, C., Griffiths, T.L., & Goodman, N. (2011). How to grow a mind: Statistics, structure, and abstraction. *Science*, Vol. 331, pp. 1279–1285; DOI: 10.1126/science.1192788.

Tversky, R., & Kahneman, D. (1983). Extensional versus intuitive reasoning: The conjunctive fallacy in probability judgment. *Psychological Review*, Vol. 90, pp. 293–315; DOI: 10.1037/0033-295X.90.4.293

Vaidman, L. (2008). Evolution stopped in its tracks. *Nature*, Vol. 45, No. 10, pp. 137–138; DOI: 10.1038/451137a.

von Neumann, J. (1955). *Mathematical Foundations of Quantum Mechanics*. Princeton University Press, Princeton.

Wheeler, J.A., & Zurek, W. H. (1983). *Quantum Theory and Measurement*. Princeton University Press, Princeton.

Yearsley, J.M., & Busemeyer, J.R. (2016). Quantum cognition and decision theories: A tutorial. *Journal of Mathematical Psychology*, Vol. 74, pp. 99–116; DOI: 10.1016/j.jmp.2015.11.005.

Yearsley, J.M., & Pothos, E.M. (2016). Zeno's paradox in decision – making. *Proceedings of the Royal Society B*, Vol. 283; DOI: 10.1098/rspb.2016.0291.

6

SUPERPOSITION AND VERACITY ASSESSMENT

Statement validity assessment

Ability of legal actors to determine the veracity of witnesses' and suspects' statements is essential for both investigations and court proceedings. However, previous research has demonstrated that when non-specialized deception detection methods are used, the overall accuracy of professional criminal investigators, prosecutors, judges, and average individuals' judgements of truthfulness does not exceed 55% (Oberlader et al., 2016; Vrij, 2008; Wojciechowski, 2016). Therefore, scientists have been developing unique methods to assist individuals in the legal system in making more precise determinations as to whether something is real or false. In the 1950s, forensic psychologists began searching for methods of evaluating testimony, including procedures for assessing the probative value of the evidence of minor witnesses, in order to solve a significant issue in the practice of law. German psychologists, including Udo Undeutsch, Friedrich Arntzten, Max Steller and Gunter Koehnken, as well as the Swedish forensic psychologist Arne Trankell, established psychologically based methods for the effective and thorough evaluation of testimonial evidence (Wojciechowski, 2014, 2016). The systematic description of testimony using psychological content criteria is founded on the belief that testimony based on direct experience of the witness differs numerically and qualitatively from evidence based on observation of real occurrences. Certain qualities separate genuine testimony from testimony that has been purposefully fabricated. Udo Undeutsch was the first to assume that testimony based on the personal experience of the witness would differ in form and content from insincere or false testimony. Undeutsch hypothesis forms the theoretical basis of content analysis, according to which content and form differences can

DOI: 10.4324/9781003402480-6

be discerned between insincere and sincere testimony (Steller & Koehnken, 1989; Wojciechowski, 2012, 2016). Raters assess the credibility of witness testimonies using a scale, form an opinion on each of the content criteria and take into account the overall score comprising the sum of the points obtained by a given testimony, its overall characteristics. They then classify the witnesses' testimonies into one of two categories – those based on direct, personal observations and those with sources other than observations and actual memories.

About a dozen methods for evaluating testimony and several hundred reliability criteria are distinguished in the psychological literature (Vrij, 2008). In multiple countries, forensic psychologists and attorneys employ content analysis to assess the testimony of minor witnesses in sexual offence trials. Experts are searching for ways to apply content analysis to processes involving also other types of crimes and to evaluate the testimony of adult witnesses and suspects are increasing (Granhag & Stroemwall, 2004; Vrij, 2009, 2003; Wojciechowski et al., 2018).

The best known and internationally widespread technique for assessing witness' statements is the statement validity assessment (SVA) by Max Steller and Guenther Koehnken (1989; Koehnken, 2004). SVA is a comprehensive technique for developing and testing hypotheses on the source of a given statement. There are nine diagnostic steps of SVA. The main stage of the assessment is the examination of the presence of 19 CBCA criteria in an interview transcript (Koehnken, 2004; Volbert & Steller, 2014). It is assumed that each of the CBCA characteristics should be more present in truthful than in false witnesses' accounts. According to the assumption on which the selection of the 19 CBCA criteria was founded, such criteria are suggestive of authentic experiences because they are frequently impossible to create (Steller & Koehnken, 1989; Volbert & Steller, 2014). Additionally, a Validity Checklist has been developed to explore aspects other than truthfulness that may affect the veracity of the statement and to investigate alternative interpretations of CBCA results (Koehnken, 2004) (Table 6.1).

More than 70 empirical investigations and few meta-analyses on SVA have been published to date (Amado et al., 2015, 2016; Bond & Lee, 2005; Hauch et al., 2017; Masip et al., 2005; Vrij, 2005, 2008, 2015; Willen & Stroemwall, 2012). These studies suggest that SVA analyses can be used for lie detection, as truth-tellers tend to receive significantly higher total scores than liars (Hauch et al., 2017; Vrij, 2008) and positive impact sizes have been reported for each credibility criterion (Volbert & Steller, 2014). The studies demonstrate that SVA is significantly more accurate than veracity assessment without specialized tools (Akehurst et al., 2015; Bond & DePaulo, 2006; Vrij, 2015), and evaluators employing the SVA criteria have a higher hit rate than clinical psychologists assessing the veracity of clients' accounts of physical and psychological symptoms in legal settings (Pezdek et al., 2004; Porter & Yuille, 1995, 1996). Laboratory experiments demonstrated that 71% of the truths and 71% of the

TABLE 6.1 Statement validity assessment content criteria

SVA criteria		
Label		*Description*
Criteria—based content analysis		
1	*Logical structure*	statement contains no contradictions or logical inconsistencies
2	*Unstructured production*	suspect reports various elements of an incident in an unsystematic, chronologically disorganized manner
3	*Quantity of details*	the event, the location and surroundings, and the people involved are described in great detail
4	*Contextual embedding*	the event in question is described as related to particular locations, time schedules, personal relationships
5	*Descriptions of interactions*	events are characterized by a sequence of actions and reactions
6	*Reproduction of conversation*	suspect reproduces conversations between different persons, using particular speech behaviour, vocabulary, etc. of those persons
7	*Unexpected complications during the incident*	the course of events is interrupted by unexpected complications and obstacles
8	*Unusual details*	refers to particular elements or details of a statement which are unexpected, surprising, e.g., odd details
9	*Superfluous details*	details which are not strictly necessary for the description of the incident
10	*Accurately reported details misunderstood*	details and actions are reported which are obviously not understood by the suspect, in particular their meaning and function
11	*Related external associations*	suspect describes events which refer to earlier events which are in some way related to the incriminated actions
12	*Accounts of subjective mental state*	the descriptions of suspect's feelings and cognitions, their development and changes during the event
13	*Attribution of victim's mental state*	descriptions of emotions, cognitions and motivations which were attributed by the suspect to the victim
14	*Spontaneous corrections*	suspect corrects or modifies previous descriptions without having been prompted by the interviewer
15	*Admitting lack of memory*	suspect expresses concern that he or she may not remember all relevant details, that the description of particular details may be incorrect
16	*Raising doubts about one's own statement*	suspect indicates that part of his or her descriptions sounds odd, implausible, unlikely, he or she can hardly believe that this is a correct account of what had happened

(*Continued*)

TABLE 6.1 (Continued)

SVA criteria	
Label	*Description*
17 *Self—depreciation*	suspect mentions personally unfavourable, self—incriminating details
18 *Pardoning the victim*	suspect excuses the victim for his or her behaviour
19 *Details specific to a type of crime*	suspect describes elements which are typical for this type of crime but, on the other hand, are counter intuitive for the general public or discrepant to everyday knowledge or stereotypes
Validity checklist	
20 *Inappropriateness of language and knowledge*	suspect's use of language and display or knowledge beyond the normal capacity of a person his or her age or beyond the scope of what the suspect may have learnt from the incident
21 *Inappropriateness of affect*	refers to whether the affect displayed by the suspect when being interviewed is inappropriate for the suspect's experiences
22 *Susceptibility to suggestion*	refers to whether suspect demonstrates any susceptibility to suggestion during the interview
23 *Suggestive, leading or coercive questioning*	examines whether the interviewer put suggestions to the interviewee or exerted any kind of pressure
24 *Overall inadequacy of the interview*	other factors determining quality of the interview
25 *Questionable motives to report*	refers to whether the suspect may have questionable motives to report the incident in certain way
26 *Questionable context of the original disclosure or report*	refers to the origin and history of the statement, particularly the context of the first report
27 *Pressures to report falsely*	deals with the question of whether there are indications that others suggested, coached, pressured or coerced the suspect to make a false report
28 *Inconsistency with the laws of nature*	refers to describing events that are unrealistic or impossible
29 *Inconsistency with other interviewee's descriptions*	refers to the possibility that major elements in the description of the core of the event are inconsistent with or contradicted by another statement made by the suspect or somebody else
30 *Inconsistency with other evidence*	refers to the possibility that major elements in the statement are contradicted by reliable physical evidence or other concrete evidence

Source: Volbert and Steller (2014).

lies were properly categorized using CBCA evaluations (Vrij, 2008), whereas a 50% accuracy rate is expected by chance alone. In 80% of trials, the hypothesis that truth-tellers will receive considerably better total CBCA ratings than liars was supported (Vrij & Ganis, 2014). In only one of the 20 studies (5%) did truth-tellers receive lower CBCA scores than liars (Ruby & Brigham, 1998), but the protocol used in that study was significantly different from the typical CBCA procedure, including the fact that assessments were based on watching videos rather than reading transcripts. In 24 research, the ability of CBCA analyses to distinguish between liars and truth-tellers was evaluated. In these trials, the average rate of accuracy for recognizing facts and lies was 71%. In other words, there is evidence that CBCA can distinguish between truths and falsehoods effectively (Vrij, 2008; Vrij & Ganis, 2014).

Although SVA assessments are admissible in court in a number of countries, it is still unknown how accurate these assessments are because there are presently no credible data regarding the accuracy of SVA assessments in actual court cases. Studies on the reliability of particular SVA content criteria have produced contradictory results; in some studies, truthful testimony meets the same criteria more frequently than false statement, while in others, there are no differences between the two (Caso et al., 2006; Oberlader et al., 2016). However, according to a meta-analysis by Hauch and colleagues (2017) the majority of CBCA criteria may be scored with sufficient to good inter-rater reliability. There are also reasons to suspect that the application of the Validity Checklist is occasionally problematic. It is difficult, if not impossible, to determine the precise influence of numerous external factors on the scores (Vrij, 2008). The use of content analysis in practice also requires the development of decision rules – so far there is no uniform indication of when content analysis should lead to the conclusion that the testimony is true and when it should be considered to contain errors or to be the result of a conscious effort to falsify the picture of reality (Goedert et al., 2005; Hauch et al., 2017; Koehnken, 2004; Masip et al., 2005Vrij, 2009, 2008, 2015). An interesting approach to the content analysis, based on decision trees and algorithms, was proposed by Wojciechowski (2012, 2014, 2016).

Definiteness and quantum superposition

According to the classical probability theory, decisions come from definite states. In other words, a decision-maker always has a clearly defined mental image of the pertinent probabilities in mind and the decision reflects already existing state (Jaeger & Trueblood, 2019). In the context of content analysis one can say that, at any point we can objectively establish if statement is truthful or false. Additionally in the veracity assessment process, the rater can always articulate existing prior to answering the question opinion about the likelihood that the witness is correctly describing reality. Decision-making is dependent on a process of accumulating evidence that generates support for potential hypotheses. Models of this process derived from classical stochastic theories assume that information accumulates by traversing distinct levels of evidence, carving a single trajectory over time (Kvam et al., 2015). Although rater's opinion may fluctuate considering new information, but any given moment expert assessing reliability of a statement can express firm opinion. Classical view assumes that probabilistic judgement regarding veracity of a statement is present in evaluator's mind at any given moment, it can be "read out" and will be only expressed, articulated when someone asks about rater's opinion.

On the other hand, in the quantum perspective it is assumed that judgements come from unbounded states. Uncertain outcomes can cause people to remain in a "*I don't know*" state until they are forced to decide. When a firm decision is prompted, it is built upon the interaction of the decision-maker's indefinite state and the question that is posed. Up until the measurement (i.e. classification of witnesses' account as true or false), multiple judgements are possible, but each measurement yields only one. This ambiguity reflects how we make decisions on a psychological level every day. We frequently experience conflict, ambiguity or uncertainty because our attitudes and opinions are not always clear. Only when we are asked a question, we are forced to make a choice, usually pause and give a clear response after giving it some thought.

The quantum probability theory allows a person to be in an indefinite (technically dispersive) state, known as a superposition state, at any given moment. Strictly speaking, this indicates that one cannot assume that psychological states are characterized by definite values that can be recorded by a psychological measurement at each instant in time. To be in a superposition state indicates that all potential definite values within the superposition have the potential to be expressed at each moment (Heisenberg, 1958; Wang et al., 2013). A superposition state is an inherent representation of the conflict, ambiguity or uncertainty encountered by cognitive processes (Blutner et al., 2013; Brainerd et al., 2013; Wang & Busemeyer, 2013). Quantum modelling allows us to formalize the state of a cognitive system moving across time in its state space (Atmanspacher & Filk, 2013; Busemeyer et al., 2006) until a decision is made, at which point the state collapses to a definite value.

In a superposed state, all conceivable definite states have the potential to materialize, but only one will actually do so when the state is measured. According to the quantum probability theory, a system is in an undefined (technically dispersed) superposition state up until it is recorded by a measurement. A superposition state is a quantum state that is expanded as a linear combination of the eigenstates of a particular operator – a weighted sum. The concept of indefiniteness in judgements can be compared to the famous "Schrödinger's cat" thought experiment in physics, in which a cat inside a box could be either dead or alive (with a small probability of each) until we open the box and check, at which point the cat must be either dead or alive. It appears that every choice we make is a different Schrödinger's cat from a quantum perspective (Jaeger & Trueblood, 2019).

The Schrödinger equation's solution yields the time-dependent function known as the quantum state, which is the fundamental concept in quantum mechanics. Hermitian operators are used to represent observable quantities, and the outcome of measuring an observable must be one of the eigenvalues of the corresponding Hermitian operator. The outcome will unquestionably be the eigenvalue corresponding to that eigenstate if the system is in an eigenstate of the operator when it is measured. The probability of getting any specific eigenvalue, if the system is in a superposition with respect to the measurement operator, is equal to the complex square of the expansion coefficient for the corresponding eigenstate (Bowman, 2008).

Superposition can be also understood as being at the same time partly in two or more different, sometimes mutually exclusive states (Mathur & Singh, 2009). Same as thinking simultaneously that "the witness is telling the truth" and "the witness is lying." The implication of this superposition is that result of the measurement would not be something intermediate between A and B (A – zero, B – one). It will be either A or B, but which one we cannot foretell until the measurement. We can only express the probability of getting the result A or B. In other words, we expect the property of the witnesses' statement to be intermediate between the component states – truthful and false. However, probability of getting the result "truth" lies between 0 and 1, and the probability of getting "lie" also lies between 0 and 1. Sum of the two probabilities (truth vs lie) equals to one, but until measurement the system is an undefined state (Mathur & Singh, 2009).

Quantum decision models assume that evidence develops over time in a superposition state analogous to a wavelike pattern, and that judgements and decisions are made through a measurement process that creates a definite state of evidence from this indefinite state. When multiple responses (measurements) are elicited over time, interference effects should occur, according to this model (Kvam et al., 2015). As a result of this superposition state, quantum probabilities permit violations of the classical law of total probability. Prior to a response, the internal state cannot be assigned a specific value according to the quantum model. Quantum theory holds that a decision (response/measurement) gives rise to potentiality. This unique form of quantum unpredictability is the defining feature of the theory that explains the emergence of interference effects (Broekaert et al., 2020; Busemeyer et al., 2006). As a consequence, the probability of assessing a statement as false when the rater is not asked to resolve the issue of possible influence of the observation conditions on statement's veracity may differ from the probability of rating evidence from witness after analysis of the circumstances of the event (Busemeyer & Wang, 2015).

How can we determine if raters are in a superposition when evaluating the veracity of a statement and quantum probability theory should be applied to model content analysis? First indicator would be an ambiguous and uncertain state of the rater prior to decision-making that is just before an opinion regarding statement's truthfulness is expressed. The second would be a constructive influence of the measurement on the judgement, a disturbing effect of the evidence accumulation and analysis on the decision to accept a witnesses' report as true or false (Pothos & Yearsley, 2023; White et al., 2014, 2016). Quantum probability theory assumes that at any moment a system is in an indefinite superposition state until measurement is performed on the system. To be in a superposed state means that all possible definite states have the potential for being actualized, but only one of them will become actual upon measurement (Bruza et al., 2015). Hypothesis, that raters' decisions have constructive role (as opposed to assumption, that they reveal objective probative value of an evidence) will be confirmed, if there will be a statistically significant difference between content analysis conclusions of the same cases, depending on whether there was an intermediate rating or conclusion was drawn without intermediate rating. Moreover, hypothesis that veracity assessments are made in the superposition state will be confirmed, if there will be statistically significant difference of reliability judgements in the same cases, depending on order of the presentation.

Study on content analysis and raters' uncertainty

The overall purpose of the experiments conducted by Wojciechowski (2012, 2016) was to examine whether and to what extent SVA can be used to improve accuracy in assessments of the veracity of statement. In line with previous findings regarding SVA's potential in distinguishing between true and false statements it was hypothesized that: veracity assessments using SVA are significantly more accurate than veracity assessments that are not based on any of the content analysis techniques. Moreover, it was expected that truthful statements will receive significantly higher overall SVA scores than false suspect statements. Collected previously data by Wojciechowski (2012, 2016) will be used to verify if raters making decision whether a statement is true or false are in a definite state or in a superposition.

Material and methods

In the experiment transcripts of witnesses' statements received in 103 Polish criminal proceedings were used. Court files were made available by the district and regional courts of the Silesian voivodeship in response to the researchers' request. Individual files were chosen by courts' authorities and secretarial staff, researchers did not have influence on their quantity and quality. From the court cases, only the ones which ended in judgement in force and statements given during the proceedings were evaluated by both instances courts and that evaluations were foundations to conduct factual adjudication of case. Information on evidential material which was collected during the proceedings, including demonstrative evidence, photographs, medical certificates, experts' opinions, as well as significance of each evidence in terms of courts' ruling and coherence between them and offender's statement content was included in a taxonomic sheet. The false testimony came from the files of proceedings conducted for a crime under Article 233 of the Criminal Code (giving false testimony), in which the suspect's confession was obtained – confirmation that the testimony given was falsified, and the circumstances previously falsified by the witness were confirmed in his explanations. In addition, the assumption that the witness's account was falsified was supported by the fact that in the proceedings in which the witness had given false testimony, on the basis of other accumulated evidence, a verdict had been issued confirming that the witness's account had been consciously, intentionally falsified, and that the picture of objects and events was consistent with his later explanations. Statements which were recorded in protocols were transcribed and raters received them in a standardized and anonymized form.

Content analysis was conducted on the testimonies of women in 43.9% of cases and men in 56.0% of cases, ranging in age from 17 to 74 years, with an average witness age of 36 years and six months (SD = 15 years and two months). The content of the testimonies included descriptions of the witnesses' direct observations in 80.5% of the cases (witness *de visu*); 18.4% of the testimonies were from witnesses basing their accounts on information they had heard (witness *de auditu*); the remaining witnesses gave testimonies on circumstances surrounding the events, collateral (1.1%). Nearly half of the testimonies subjected to content analysis came from witnesses who were victims of a crime (45.9%). Most testimonies were consistent and coherent, with only 18.2% of the testimonies analysed showing significant changes in content. The content of all testimony given by the witness throughout the proceedings, between one and seven hearings (M = 2.51, SD = 1.19), consisted of between 141 and 3146 words (M = 1072.41, SD = 787.89). It is noteworthy that the volume of true testimonies was lower than the volume of false testimonies. Statistically significant differences were found between the volume of true testimony and the number of words used by witnesses falsifying the testimony ($F(1,612) = 50.65$, $p < 0.01$), but only 8.6% of the variance in testimony volume could be explained by the evidential value of the testimony ($\eta^2 = 0.08$) (Wojciechowski, 2012, 2016).

The SVA questionnaire was used to evaluate statements that contained a table presenting the content criteria of a given method with a brief description (description of certain criteria meaning). To evaluate the level of coherence between the statement and the given criteria, a 6-point scale was utilized.

In our study, one participant indicated that the criterion was not present or fulfilled at all, whereas six participants indicated that the criterion was present or fulfilled to a very high degree. The content analysis of statements using SVA was described in terms of all content criteria pertinent to the method. The evaluation incorporated all 30 SVA criteria. Raters evaluated each criterion separately and totalled the points awarded for each criterion. In the final section of the worksheet, there were columns for evaluating the evidential value of a statement based on its method (true/false). Alternately, the evaluator could evaluate the statement's veracity independently of the content analysis results (true/false) and add comments.

Participants

Thirty-four psychology students from the University of Silesia in Katowice, Poland, participated in the study (30 women and four men). Statement analysis was a component of the curriculum for the participants' forensic psychology course. Participants ranged in age from 21 to 33 (M = 22.92 years, SD = 2.13). Participation in this study was entirely voluntary for all participants. Informing participants that the purpose of the study is to examine the validity and precision of content analysis methods. Participants received a consent form (for participation in the study) prior to the study and a debriefing form following its conclusion. No compensation, remuneration, or credit was given to participants for their participation in the study. At the outset of the study, participants indicated the number of content analyses they were willing to conduct and were then randomly assigned to one of the groups. Most participants (N = 24) received training in SVA, while the remainder (N = 10) assessed veracity without training and without employing any content analysis technique. The participants who did receive training participated in a four-day, 26-hour statement analysis course. Course topics included definitions of truths and lies, theoretical assumptions of content analysis techniques, and exercises utilizing the SVA method for statement analysis. The instructor elaborately described and illustrated each SVA criterion, and then discussed them with the students. The training encouraged students to ask questions and participate actively. The completion of the required training was a prerequisite for analysing the suspects' statements included in the study.

Results

We have collected 159 ratings. The analysis of accuracy rates for distinguishing between truthful and false accounts revealed that both training and content analysis technique had a significant impact on accuracy. Untrained participants correctly classified 33.9% of statements, including 38.6% and 26.7% false statements. SVA was accurate in 61.8% of cases, but this technique was more accurate at detecting lies, resulting in 69.2% of correct classifications, compared to a rate of 55.7% for detecting the truth. The overall accuracy of assessing veracity with SVA was significantly higher than with untrained judgement (χ = 93.12, p<.001).

Taking into consideration SVA's assumptions we could expect that overall results will be higher for truthful statements, because it would indicate that the account is experience and memory based. Additionally, each criterion of the method will be rated higher for accounts based on personal memories, in comparison with false statements. Coders were asked to rate 30 SVA criteria on a 6-point scale (if a criterion was not present at all in the statement it received a score of 1, and if it was strongly present, it received score of six); therefore, SVA scores could range from 30 to 180. There was no statistically significant difference between the overall sum of points between the truthful and false statements ($M_{truthful}$=116.23, $SD_{truthful}$=15.90; M_{false}=111.56, SD_{false}=21.13; t(152)=1.27, p=.10).

To assess the effectiveness of the individual SVA criteria and test the significance of differences between truthful and false accounts the independent samples *t-test* across individual criteria was carried out. At least for part of the individual SVA content criteria we have observed significant differences between truths and lies, that is, for: 1. (Logical structure), 2. (Unstructured production), 3. (Quantity of details), 14. (Spontaneous corrections), 16. (Raising doubts about one's own statement), 20. (Inappropriateness of language and knowledge), 25. (Questionable motives to report), 26. (Questionable context of the original disclosure or report), 27. (Pressures to report falsely), 29. (Inconsistency with other interviewee's descriptions) and 30 (Inconsistency with other evidence) (for more please see: Wojciechowski, 2012, 2016).

SVA is a verbal veracity assessment tool used to determine credibility of witnesses' statements. It is a method similar to numerous other techniques used to collect data, measure defined qualities and quantities, guide the rater and enable accurate veracity assessment. To test, if content analysis is a process of accumulating information allowing rater to reach a definite state and a firm opinion on witnesses' account (as truthful or false), we have calculated simple moving means (rolling means) for the individual SVA criteria. Simple rolling means are unweighted means of the last n values. They are calculated by averaging previous means. For example, rolling mean for the second criteria will be equal to average of means for the first and second criteria divided by 2 (rolling M_2=($M_{criterion1}$+$M_{criterion2}$)/2), rolling mean for the fifth criterion will be calculated as sum of means for criteria from 1 to 5, divided by 5 (rolling M_5=($M_{criterion1}$+$M_{criterion2}$+ $M_{criterion3}$+$M_{criterion4}$+ $M_{criterion5}$)/5) and moving mean for all 30 criteria of the SVA will be calculated as sum of the means for each of criteria from 1 to 30 divided by 30 ($\Sigma M_{criterion1\ to\ 30}$/30). Rolling means can help to find trends that would otherwise be hard to detect. We can expect that ratings (and means) for individual criteria may vary due to numerous factors

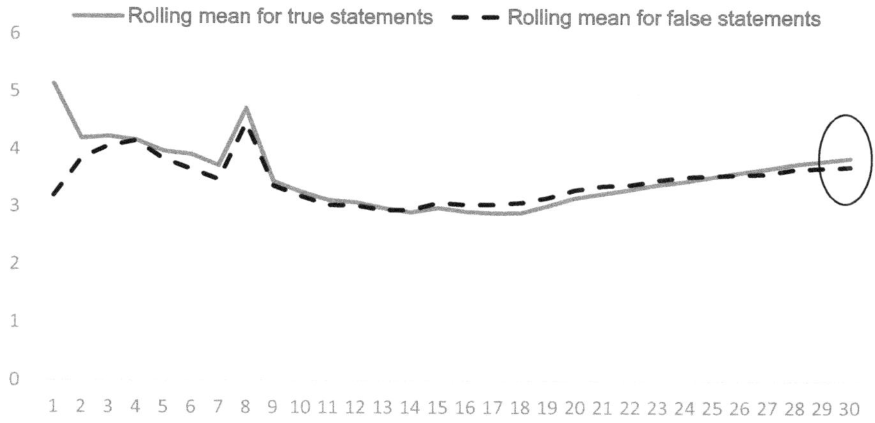

FIGURE 6.1 Rolling means for statements classified as true and false.

(i.e., compatibility of the criterion with the testimony, level of training, type of crime). However, we should be able to observe an upward trend (increasing rolling mean) for genuine statements and downward trend (decreasing rolling mean) for disingenuous reports. Collected data have been used to calculate rolling means separately for statements that were classified by the raters as true and false (please see Figure 6.1).

We can draw three conclusions from Figure 6.1: regardless of whether the evaluator finds the testimony trustworthy or believes it as untrue, almost each of the SVA criterion is assigned with the same number of points (except for criterion 1 and criterion 2); SVA does not help to build the grounds for decision and does not determine conclusion; before determining statement's veracity, those raters who label a testimony as unreliable are just as uncertain as those who label it as credible.

Content analysis and constructive influence of measurement

The second study was conducted to explore the evidence for measurement impacts (evaluation biases) on forensic judgement (concerning veracity of a statement). Some data suggest that making a decision can modify the mental states involved in the decision (Pothos & Yearsley, 2023). Quantum probability theory assumes that at any moment a system is in an indefinite superposition state until measurement is performed on the system. To be in a superposed state means that all possible definite states have the potential for being actualized, but only one of them will become actual upon measurement (Bruza et al., 2015). To test predictions of the quantum probability theory and establish if raters are in a superposition when expressing opinion on statement's veracity, we have collected three types of SVA ratings: in the first scenario participants conducted analysis of all individual 30 SVA content criteria and decision on statement's veracity; in the second scenario, participants contacted Criteria Based Content Analysis with the use of 19 content criteria, made intermediate rating on testimony's truthfulness, then analysis of the 11 Validity Checklist criteria and then the final rating; in the third scenario participants began with the Validity Checklist, expressed intermediate rating on veracity, conducted Criteria Based Content Analysis, and made final rating.

Following the QPT predictions we assume, that the final ratings are going to be different depending on order of presentation (all SVA criteria together, CBCA first – intermediate rating - VC second – final rating, VC first – intermediate rating – CBCA second – final rating). CBCA is more focused on qualities and quantities of the statement, confirmation of veracity and VC is used to confront witnesses' report with outside resources and knowledge about the events. We also predict, following works by White et al. (2014, 2016, 2020) that there will be difference in ratings of a positively valenced stimulus (statement is true), depending on whether a previous oppositely valenced stimulus (statement is false) was concluded in the intermediate rating or not. We also expect analogical difference in ratings of a negatively valenced stimulus, if previous positively valenced stimulus was rated.

We assume that the participant's initial cognitive is set by the first stimulus in sequence. Following the intermediate judgement of the first stimulus, the cognitive state is changed to being either positive or negative (truth or lie). Having made the intermediate rating, when the second oppositely valenced stimuli are presented, it is evaluated from the perspective of a different cognitive state, than it would have been without an intermediate rating. Since the second stimuli is opposite in valence to the first, there is a greater contrast in the impression made by the second stimuli. Without the intermediate rating, the differences between the stimuli concern not only aspects of their quality, but also aspects that are not related to judgement, so the affective state between the first and second stimuli is not pronounced and not established. It is in this way that the quantum probability model prediction arises, that the intermediate rating increases the contrast between the two stimuli.

In the model, change of rating is represented by a collapse of the state vector on either positive (true) or negative (false) judgement ray. Cognitive processes are based on multi-dimensional vector spaces (called Hilbert spaces) in which the so-called state vector is assumed to correspond to the relevant cognitive state (Busemeyer et al. 2015; Busemeyer & Wang, 2018a, 2018b). In such spaces, different subspaces represent different possibilities for the cognitive state vector. An important operation in quantum probability theory is that of a projection: a projection takes a vector and "lays it down" on a particular subspace. Squared length of a projection, along a subspace, determines the probability that the corresponding possibility is true (of the system represented by the state vector). If we ask subject to rate probability that statement is true, the squared length of the projection in the true subspace is the probability that the answer will be yes. If the answer is yes, then the cognitive state vector changes and is now a vector (of length one), along the projection. The greater projection (overlap) between the state vector and a subspace, the higher the probability. The higher consistency between two possibilities, the smaller angle between the corresponding subspaces, since it is easier to project from one subspace to other. Mutually exclusive possibilities, such as truth and lie, imply a 90 degree angle between the subspaces – when the state vector is along the subspace "truthful," the projection to the "false" subspace is zero (when probability of "truth" = 1, then probability of "guilty" = 0).

In quantum probability theory dynamic processes can be modelled as rotations of the state vector (Wang & Busemeyer, 2013). Information received by the subject is represented by rotation towards the proper subspace, affecting the probability. For example, suppose that the rater receives information that would indicate that witness is hiding information (his DNA was found at the crime scene and he did not inform about his presence at the place), then the cognitive state vector would be rotated toward the "false" ray, and probability that the rater will find the statement as unconvincing will increase. When the witness explain that suspect has stolen his bag with a hairbrush, rater's cognitive state vector would be rotated toward the "truth" ray, and the likelihood of assessing the statement as false will decrease.

We develop a prediction from the quantum probability theory that unless the cognitive state vector is wholly within positive or negative affect ray, then the observer is in a superposition regarding his/her judgement of the statement. Superposition means, that the state is not consistent with either evaluation, prior to making a judgement. And once a judgement is made, the cognitive state vector transitions the ray consistent with the result of the judgement, and judgements consistently lead to the expected transitions.

In the Validity Checklist–Content Veracity Assessment order, contrast between the second stage of content analysis and the cognitive state prior to its introduction will determine how negatively the second part of the case description will be rated. Without the intermediate veracity rating, we assume that the second part of description will be rated positively, so the cognitive state changes to correspond to positive judgement. Then, the perception of the qualities indicating deception will depend on the contrast between CBCA and the Validity Checklist judgement. In the double rating conditions, the observer forces a transition of the cognitive state vector to the positive judgement ray (truth) or the negative judgement ray (guilty). This is an additional change to the state vector, which compounds with the change from introducing the second set of content criteria. Based on model proposed by White and colleagues (2014, 2016) we predict, that when CBCA criteria are rated at first and intermediate rating is expressed, after rating the Validity Checklist criteria cognitive state vector will be closer to the false ray, increasing probability, that in the final rating statement will be classified as false, and exactly vice versa for the VC–CBCA condition. For the CBCA–VC condition, an intermediate rating of veracity for the CBCA criteria will result in a more negative rating for the CBCA than if the statement would be rated after complete SVA. The prediction for the VC–CBCA condition is reversed, in that the intermediate judgement will result in assessment of the statement as more truthful. In both cases, the intermediate evaluation is predicted to increase the difference in the perception of validity at the second stage of content analysis, resulting in more polarized rating of the first and second group of the veracity of the statement.

Hypothesis, that when making legal decisions raters are in the superposition state and content analysis has a constructive role will be confirmed, if there will be statistically significant difference of the SVA ratings conclusions in the same cases, depending on whether there was an intermediate rating or evaluator's opinion was expressed without intermediate rating. If in identical task the intermediate judgement will influence somehow the cognitive state and outcome of the veracity assessment process, the prediction that the measurement of internal state can have a constructive role will be confirmed. Moreover, hypothesis that legal decisions are made in the superposition state will be confirmed, if there will be statistically significant difference of ratings for the same cases, depending on order of the presentation, that is whether participants rated CBCA criteria first or started with the Validity Checklist and then conducted the Criteria Based Content Analysis.

Material and methods

In the second experiment, we have used the same transcripts of witnesses' statements, as in the first study and the same classification and selection procedure. Content analysis was conducted on 24 different testimonies, 45.8% from women and 54.2% from men, ranging in age from 17 to 59 years, with an average witness age of 28 years and 9 months (SD = 9 years and 2 months). The content of the testimonies included descriptions of the witnesses' direct observations in 87.3% of the cases (witness *de visu*); the rest of the testimonies were from witnesses basing their accounts on information they had heard (witness *de auditu*).

Same SVA questionnaire as in the first study was used. The evaluation incorporated all 30 SVA criteria, that is 19 of the Criteria Based Content Analysis and 11 of the Validity Checklist. Raters evaluated each criterion separately on a 6-point scale. In the first scenario participants evaluated all SVA criteria and in the final section of the worksheet expressed their opinion, whether the statement is true or false. In the second scenario, raters evaluated only CBCA or VC criteria, expressed intermediate opinion, then rated remaining SVA's content criteria (VC or CBCA, respectively) and finally evaluated statement's veracity. Altogether we have collected 151 ratings, 68 overall ratings (first all SVA criteria evaluated and then statements' veracity was assessed) which represents 45.0% of the data, 40 CBCA–VC ratings (CBCA assessment – intermediate judgement – VC assessment – final judgement) which is 26.5% of the collected cases and 43 VC–CBCA ratings (28.5%). True statements were evaluated 85 times (56.3%) and false statements were rated 66 times (43,7%).

Participants

Thirty-four psychology students from the University of Silesia in Katowice, Poland, participated in the study (26 women and 8 men). Statement analysis was a component of the curriculum for the participants' forensic psychology course. Participants ranged in age from 22 to 30 (M = 23.73 years, SD = 1.29). Participation in this study was entirely voluntary for all participants. Participants were informed that the purpose of the study is to examine the validity and reliability of content analysis methods. Participants received a consent form (for participation in the study) prior to the study and a debriefing form following its conclusion. No compensation, remuneration, or credit was given to participants for their participation in the study. At the outset of the study, participants indicated the number of content analyses they were willing to conduct and were then randomly assigned to one of the groups. All participants received training in SVA (a five-day, 25-hour statement analysis course). Course topics included definitions of truths and lies, theoretical assumptions of content analysis techniques, and exercises utilizing the SVA method for statement analysis. The instructor elaborately described and illustrated each SVA criterion, and then discussed them with the students. The training encouraged students to ask questions and participate actively. The completion of the required training was a prerequisite for analysing the suspects' statements included in the study. Cases and assessment variants (SVA, CBCA–VC, VC–CBCA) were randomly assigned to the raters.

Results

We have collected 151 ratings, 85 assessments of truthful (56.3%) and 66 of false statements (43.7%). Before analysis of the main hypothesis, we have assessed the value of statements analysed by the participants by comparing the SVA, CBCA and VC ratings for truthful and false witnesses' accounts. Coders were instructed to grade 30 SVA criteria on a 6-point scale (if a criterion was not present in the statement at all, it was given a score of 1, and if it was strongly present, it was given a score of 6), therefore SVA scores might vary from 30 to 180, CBCA scores could vary from 19 to 114 and VC scores from 11 to 66. Taking SVA's assumptions into account, we could anticipate that overall SVA rating, as well as CBCA and VC ratings will be greater for accurate assertions, as this would imply that the account is founded on experience and memory. Comparison of the means confirmed the assumptions, we have observed significant differences in the total SVA scores ($M_{truthful}$=118.96, $SD_{truthful}$=13.65; M_{false}=104.71, SD_{false}=12.53; $t(150)$=6.60, p<.05), CBCA scores ($M_{truthful}$=66.76, $SD_{truthful}$=11.11; M_{false}=54.15, SD_{false}=10.68; $t(150)$=7.02, p<.01) as well as sum of point assigned to Validity Checklist ($M_{truthful}$=52.20, $SD_{truthful}$=5.37; M_{false}=50.56, SD_{false}=6.09; $t(150)$=1.75, p<.05).

To test the main hypotheses and verify if measurement has effect on the result, we have compared overall sum of SVA points assigned to true and false statements analysed with or without the intermediate rating of veracity. Two-way ANOVA revealed that there are significant differences between the overall sum of SVA points assigned to statements, depending on their veracity and the manner in which the evaluation was conducted. Sum of points was higher for true statements, when raters conducted full SVA, in comparison to ratings with intermediate assessment ($M_{TRUEoverall}$=116.62, $SE_{TRUEoverall}$=1.90; $M_{TRUEintermediate}$=113.56, $SE_{TRUEintermediate}$=1.81; p<.001), and lower for false statements when no intermediate rating was introduced ($M_{FALSEoverall}$=99.00, $SE_{FALSEoverall}$=4.10; $M_{FALSEinttermediate}$=107.62, $SE_{FALSEintermediate}$=3.10, p<.01; $F(1,150)$=4.20, p<.05). There were also significant, analogous differences in the CBCA ratings, depending on statements' veracity and type of measurement (overall vs intermediate); true statements received more points when raters conducted full SVA, in comparison to ratings with intermediate judgement ($M_{TRUEoverall}$=64.43, $SE_{TRUEoverall}$=1.57; $M_{TRUEinttermediate}$=62.40, $SE_{TRUEintermediate}$=1.4; p<.01), and lower for false statements when no intermediate rating was introduced ($M_{FALSEoverall}$=47.75, $SE_{FALSEoverall}$=3.40; $M_{FALSEinttermediate}$=57.09, $SE_{FALSEintermediate}$=2.57, p<.001; $F(1,150)$=21.02, p<.001). When it comes to results of the measurement with the use of Validity Checklist criteria, there were no significant differences between intermediate and overall ratings ($F(1,150)$=.477, p=.49)

Further analyses were focused on order effects in the intermediate rating scenario – search for differences between CBCA and VC ratings depending on the sequence in which they were carried out (CBCA first and then VC vs VC and then CBCA). The two-way interaction for assessments of the true and false statements between the CBCA–VC and VC–CBCA orders was significant ($F(1,82)$ = 20.98, p<.001). Raters assigned significantly more points to the true statements when Criteria Based Content Analysis was conducted as first, in comparison to true statements rated with the CBCA criteria after the Validity Checklist ($M_{TrueCBCAfirst}$=69.80, $SE_{TrueCBCAfirst}$=2.85; $M_{TrueCBCAsecond}$=64.13, $SE_{TrueCBCAsecond}$=1.19; p<.05). False statements received more CBCA points when raters began their analysis with the Validity Checklist and less points when the VC opened the assessment ($M_{FalseCBCAfirst}$=53.80, $SE_{FalseCBCAfirst}$=1.21;

$M_{FalseCBCAsecond}$=56.00, $SE_{FalseCBCAsecond}$=1.33; $p<.05$). Order of assessment with the Validity Checklist criteria (before or after CBCA) did not have significant influence on the results ($F(1,82)=.37$, $p=.54$).

At the last stage of the analysis, we compared final CBCA and VC ratings, depending on intermediate classification of the statement as truthful or false. It is by analysing double interaction that we can assess evidence for constructive role of the measurement, that is: by comparison of CBCA/VC ratings after intermediate veracity assessment and CBCA/VC ratings before preliminary rater's opinion that a statement is true or false. There was a significant interaction effect of rater's opinion about statement's veracity (true vs false) on CBCA ratings (initial vs secondary) ($F(1, 76)=13{,}49$, $p<.001$). When Criteria Based Content Analysis was conducted prior to intermediate classification of the statement as credible, CBCA scores were higher ($M_{CredibleCBCAfirst}$=76.13, $SE_{CredibleCBCAfirst}$=2.57) than when Criteria Based Content Analysis has led to finding the testimony credible, but followed Validity Checklist ($M_{CredibleCBCAsecond}$=62.31, $SE_{TrueCBCAsecond}$=2.01; $p<.05$); unreliable statements received less points when CBCA preceded Validity Checklist in comparison to false statements assessed with the CBCA after Validity Checklist ($M_{UnreliableCBCAfirst}$=52.71, $SE_{UnreliableCBCAfirst}$=2.82; $M_{UnreliableCBCAsecond}$=57.07, $SE_{UnreliableCBCAsecond}$=1.40; $p<.05$). For the Validity Checklist, there was also an interaction effect ($F(1,75)=4.63$, $p<.05$): credible accounts received more VC points at the first stage of assessment, when Validity Checklist preceded CBCA ($M_{CredibleVCfirst}$=53.35, $SE_{CredibleVCfirst}$=1.38) than when final judgement followed the assessment with VC criteria ($M_{CredibleVCsecond}$=50.62, $SE_{TrueVCsecond}$=1.01; $p<.05$); for the unreliable testimonies, there was an opposite relationship – when VC was the first step of analysis false statements received significantly less points, than when the Validity Checklist was used after CBCA ($M_{UnreliableVCfirst}$=48.19, $SE_{CredibleVCfirst}$=1.52; $M_{UnreliableVCsecond}$=51.24, $SE_{UnreliableVCsecond}$=1.30; $p<.05$).

Collected data indicate evaluation bias, order effects and constructive role of measurement. Results confirm that raters analysing veracity of a statement are in a superposition. Conclusion following assessment with the content criteria do not reflect prior state (objective value of the statement) but are a result of a process of decision-making affected by the measurement.

Constructive impacts and the collapse principle have been observed in numerous studies on application of the quantum probability theory to cognition (Gloeckner et al., 2010; Pothos & Yearsley, 2023). The order of evidence influences the final judgement, and people frequently change the subjective cue validity associated with a decision in a direction that is consistent with the decision (Hogarth & Einhorn, 1992; Trueblood & Busemeyer, 2011). Order effects occur in a range of inference tasks, including assessing the probability of selecting balls from urns and determining a defendant's guilt or innocence in a simulated trial (Furnham, 1986; Walker et al., 1972).

Quantum probability has also been used to explain incompatibility and interference in other spheres of human reasoning (Aerts & Aerts, 1995; Khrennikov, 2014; White et al., 2014, 2016, 2020; Wojciechowski et al., 2022). White et al. (2014, 2020) applied pairs of stimuli with significant positive or negative valence, with the first stimulus being scored just 50% of the time. Evaluation bias has been proven multiple times and quantum probability assumption that rating the first stimulus would intensify the evaluation of the second stimulus (i.e., make it more positive if it was already positive) was confirmed. Kvam and colleagues (2015) have built and experimentally verified quantum random – walk model of evidence accumulation during judgement and decision-making. They have demonstrated substantial empirical evidence for the quantum prediction that choices interfere with subsequent confidence assessments and proved that quantum random walk is a feasible alternative to the Markov random walk model for quantitatively fitting choice and confidence distributions. Wang and Busemeyer (2015) also empirically examined quantum and Markov random walk models in an experiment in which participants evaluated the efficacy of public health service announcements from either their own or another person's perspective. The model comparison in light of the results significantly supported the quantum random walk model over the Markov model.

Trueblood and Busemeyer (2011) developed a new experiment involving extreme evidence and fitted the quantum model to data collected in a medical diagnostic task and a jury decision-making task. The results of this new experiment imply that the belief-adjustment model is limited in its ability to account for tasks containing extreme evidence, whereas the quantum inference model is not. Ultimately, they argue that the quantum model provides a previously impossible cohesive explanation for order effects. Study conducted by Wojciechowski et al. (2022) explored the evidence for positive impacts (evaluation biases) on clinical judgement (concerning eating disorders). Collected data also confirmed constructive processes in the judgements of our participants.

Results of our study motivate fitting the quantum model of White et al. (2014, 2020), the fundamental quantum operationalization of evaluation bias. In this model a two-dimensional subspace is assumed, with separate one-dimensional subspaces for truth (testimony based on the personal experience of a witness) and lie (insincere or false account). We can consider the example of SVA – CBCA rating. A participant would initially be exposed to information indicating that witness may present false information; hence, the initial mental state vector is positioned close to the lie subspace. Probability that statement will be assessed as false depends on the angle between vector representing state of the rater and the lie subspace. It is computed as the squared length of the projection of the state vector on the subspace corresponding to truth or lie. Order effects and evaluation biases are modelled with a quantum-like model assuming, that when rater focuses attention on CBCA criteria indicating that report is based on memory, the vector is rotated (moved) towards the truth ray. In the single rating condition, there is only one assessment, vector representing decision-maker's state rotates until the final decision is made. In the intermediate rating scenario there are two judgements, first one should result in a projection of the state vector along the lie ray following the initial characteristics of the account. Introduction of the CBCA content criteria indicating truthfulness would then cause a rotation of the state vector, but movement of the vector will begin from the projection on the lie subspace. In both instances, the probability of truth is computed by projecting along the healthy ray, but since evolution of the system and rotation of the state vector were different – evaluation bias can not only be described but also explained.

References

Aerts, D., & Aerts, S. (1995). Applications of quantum statistics in psychological studies of decision processes. *Foundations of Science,* Vol. 1, pp. 85–97.

Akehurst, L., Easton, S., Fuller, E., Drane, G., Kuzmin, K., & Litchfield, S. (2015). An evaluation of a new tool to aid judgments of credibility in the medico—legal settings. *Legal and Criminological Psychology*, Vol. 22, pp. 22–46; DOI: 10.1111/lcrp.12079.

Amado, B.G., Arce, R., & Fariña, F. (2015). Undeutsch hypothesis and criteria based content analysis: A meta— analytic review. *European Journal of Psychology Appl to Legal Context*, Vol. 7, pp. 3–12; DOI: 10.1016/j.ijchp.2016.

Amado, B.G., Arce, R., Fariña, F., & Vilariño, M. (2016). Criteria—Based content analysis (CBCA) reality criteria in adults: A meta—analytic review. *International Journal of Clinical and Health Psychology*, Vol. 16, pp. 201–210; DOI: 10.1016/j.ijchp.2016.01.002.

Blutner, R., Pothos, E.M., & Bruza, P. (2013). A quantum probability perspective on borderline vagueness. *Topics in Cognitive Sciences*, Vol. 5, pp. 711–736; DOI: 10.1111/tops.12041.

Bond, C.F., & DePaulo, B.M. (2006). Accuracy of deception judgements. *Personality and Social Psychology Review*, Vol. 10, pp. 214–234; DOI: 10.1207/s15327957pspr1003_2.

Bond, G.D., & Lee, A.Y. (2005). Language of lies in prison: Linguistic classification of prisoners' truthful and deceptive natural language. *Applied Cognitive Psychology*, Vol. 19, pp. 313–329; DOI: 10.1002/acp.1087.

Bowman, G.E. (2008). *Essential Quantum Mechanics.* Oxford University Press, Oxford.

Brainerd, Ch.J., Wang, Z., & Reyna, V.F. (2013). Superposition of episodic memories: Overdistribution and quantum models. *Topics in Cognitive Sciences*, Vol. 5, pp. 773–799; DOI: 10.1111/tops.12039.

Broekaert, J.B., Busemeyer, J.R., & Pothos, E.M. (2020). The disjunction effect in two-stage simulated gambles. An experimental study and comparison of a heuristic logistic, Markov and quantum-like model. *Cognitive Psychology*, Vol. 117, p. 101262; DOI: 10.1016/j.cogpsych.2019.101262.

Bruza, P.D., Wang, Z., & Busemeyer, J.R. (2015). Quantum cognition: A new theoretical approach to psychology. *Trends in Cognitive Sciences*, Vol. 19, No. 7, pp. 383–393; DOI: 10.1016/j.tics.2015.05.001

Busemeyer, J.R., Wang, Z., & Townsend, J.T. (2006). Quantum dynamics of human decision-making. *Journal of Mathematical Psychology*, Vol. 50, pp. 220–241; DOI: 10.1016/j.jmp.2006.01.003.

Busemeyer, J.R., & Wang, Z. (2015). What is quantum cognition, and how is it applied to psychology? *Current Directions in Psychological Science*, Vol. 24, No. 3, pp. 163–169; DOI: 10.1177/0963721414568663.

Busemeyer, J.R., Wang, Z., & Pothos, E. (2015). Quantum models of cognition and decision, [in:] *Oxford Handbook of Computational and Mathematical Psychology*, [eds.] Busemeyer, J.R., Wang, Z., Townsend, J.T., & Eidels, A., Oxford University Press, New York, pp. 369–389.

Busemeyer, J.R., & Wang, Z. (2018a). Hilbert space multidimensional theory. *Psychological Review*, Vol. 125, No. 4, pp. 572–591; DOI: 10.1037/rev0000106.

Busemeyer, J.R., & Wang, Z. (2018b). Data fusion using Hilbert space multi – dimensional models. *Theoretical Computer Science*, Vol. 752, pp. 41–55; DOI: 10.1016/j.tcs.2017.12.007

Caso, L., Vrij, A., Mann, S., & De Leo, G. (2006). Deceptive responses: The impact of verbal and non-verbal countermeasures. *Legal and Criminological Psychology*, Vol. 11, pp. 99–111; DOI: 10.1348/135532505X49936.

Furnham, A. (1986). The roubustness of the recency effect: Studies using legal evidence. *The Journal of General Psychology*, Vol. 113, No. 4, pp. 351–357; DOI: 10.1080/00221309.9711045.

Gloeckner, A., Betsch, T., & Schindler, N. (2010). Coherence shifts in probabilistic inference tasks. *Journal of Behavioral Decision Making*, Vol. 23, No. 5, pp. 439–462; DOI: 10.1002/bdm.668.

Goedert, H.W., Gamer, M., Rill, H.G., & Vossel, G. (2005). Statement validity assessment: Interrater reliability of criteriabased content analysis in the mock-crime paradigm. *Legal and Criminological Psychology*, Vol. 10, pp. 225–245.

Hauch, V., Sporer, S.L., Masip, J., & Blandon-Gitlin, I. (2017). Can credibility criteria be assessed reliably? A meta— Analysis of criteria—Based content analysis. *Psychological Assessment*, Vol. 29, pp. 819–834; DOI: 10.1037/pas0000426.

Heisenberg, W. (1958). *Physics and Philosophy: The Revolution in Modern Science, Role of Modern Physics in Human Thinking*. Harper & Row, New York.

Hogarth, R.M., & Einhorn, H.J. (1992). Order effects in belief updating: The belief-adjustment model. *Cognitive Psychology*, Vol. 24, pp. 1–55; DOI: 10.1016/0010–0285(92)90002-J.

Jaeger, Ch.B., & Trueblood, J.S. (2019). Thinking quantum: A new perspective on decisionmaking in law. *Florida State University Law Review*, Vol. 46, No. 4, pp. 733–806.

Khrennikov, A. (2014). *Beyond Quantum*. Taylor & Francis Group, New York.

Koehnken, G. (2004). Statement validity analysis and the 'detection of the truth, [in:] *Detection of Deception in Forensic Contexts,* [eds.] Granhag, P.A., & Stroemwall, L.A., Cambridge University Press, Cambridge, pp. 41–63.

Kvam, P.D., Pleskac, T.J., Yu, S., & Busemeyer, J.R. (2015). Interference effects of choice on confidence: Quantum characteristics of evidence accumulation. *PNAS*, Vol. 112, No. 34, pp. 10645–10650; DOI: 10.1073/pnas.1500688112.

Masip, J., Sporer, S.L., Garrido, E., & Herrero, C. (2005). The detection of deception with the reality monitoring approach: A review of the empirical evidence. *Psychology, Crime and Law*, Vol. 11, pp. 99–122; DOI: 10.1080/10683160410001726356.

Mathur, V.S., & Singh, S. (2009). *Concepts in Quantum Mechanics.* CRC Press, Boca Raton.

Oberlader, V.A., Naefgen, C., Koppehele-Gossel, J., Quinten, L., Banse, R., & Schmidt, A.F. (2016). Validity of content—Based techniques to distinguish true and fabricated statements: a meta-analysis. *Law and Human Behaviour*, Vol. 40, pp. 440–457; DOI: 10.1037/lhb0000193.

Pezdek, K., Morrow, A., Blandon-Gitlin, I., Goodman, G.S., Quas, J.A., Saywitz, K.J., & Brodie, L. (2004). Detecting deception in children: Event familiarity affects criterion-based content analysis ratings. *Journal of Applied Psychology*, Vol. 89, No. 1, pp. 119–126; DOI: 10.1037/0021–9010.89.1.119.

Porter, S., & Yuille, J.C. (1995). Credibility assessment of criminal suspects through statement analysis. *Psychology, Crime and Law*, Vol. 1, pp. 319–331.

Porter, S., & Yuille, J.C. (1996). The language of deceit: An investigation of the verbal clues to deception in the interrogation context. *Law and Human Behaviour*, Vol. 20, pp. 443–458.

Pothos, E.M. & Yearsley, J.M. (2023). Quantum cognition: Quo vadis? [in:] *The Quantum – Like Revolution,* [eds.] Plotnitsky, A., & Haven, E., Springer Nature, Cham, pp. 281–297; DOI: 10.1007/978-3-031-12986-5_14.

Ruby, C.L. & Brigham, J.C. (1998). Can criteria-based content analysis distinguish between true and false statements of African-American speakers? *Law and Human Behavior*, Vol. 22, No. 4, pp. 369–388; DOI: 10.1023/A:1025766825429.

Steller, M., & Koehnken, G. (1989). Criteria-based content analysis, [in:] *Psychological Methods in Criminal Investigation and Evidence,* [ed.] Raskin, D.C., Springer, New York, pp. 217–245.

Trueblood, J.S., & Busemeyer, J.R. (2011). A quantum probability account of order effects in inference. *Cognitive Science*, Vol. 11, pp. 1518–1552; DOI: 10.1111/j.1551–6709.2011.01197.x.

Volbert, R., & Steller, M. (2014). Is this testimony truthful, fabricated, or based on false memory? Credibility assessment 25 years after Steller and Koehnken. *European Psychology*, Vol. 19, pp. 207–220; DOI: 10.1027/1016–9040/a000200.

Vrij, A. (2005). Criteria-based content analysis: A qualitative review of the first 37 studies. *Psychology, Public Policy and Law*, Vol. 11, pp. 3–41; DOI: 10.1037/1076–8971.11.1.3.

Vrij, A. (2008). *Detecting Lies and Deceit. Pitfalls and Opportunities.* Wiley, Chichester.

Vrij, A. (2015). Verbal lie detection tools: Statement validity analysis, reality monitoring and scientific content analysis, [In:] *Detecting Deception. Current Challenges and Cognitive Approaches,* [eds.] Granhag, P.A., Vrij, A., & Verschuere, B., Wiley, Chichester, pp. 3–36.

Vrij, A., Akehurst, L., Soukara, S., & Bull, R. (2004a). Detecting deceit via analyses of verbal and nonverbal behavior in children and adults. *Human Communication Research*, Vol. 30, pp. 8–41; DOI: 10.1111/j.1468–2958.2004. tb00723.x.

Vrij, A., Akehurst, L., Soukara, S., & Bull, R. (2004b). Let me inform you how to tell a convincing story: CBCA and reality monitoring scores as a function of age, coaching, and deception. *Canadian Journal of Behavioural Sciences*, Vol. 36, No. 2, pp. 113–126; DOI: 10.1037/h0087222.

Vrij, A., & Mann, S. (2006). Criteria-based content analysis: An empirical test of its underlying processes. *Psychology, Crime and Law*, Vol. 12, pp. 337–349; DOI: 10.1080/10683160500129007

Walker, L., Thibaut, J., & Andreoli, V. (1972). Order of presentation at trial. *Yale Law Journal*, Vol. 82, pp. 216–226.

Wang, Z., Busemeyer, J.R., Atmanspacher, H., & Pothos, E.M. (2013). The potential of using quantum theory to build models of cognition. *Topics in Cognitive Sciences*, Vol. 5, pp. 672–688; DOI: 10.1111/tops.12043.

Wang, Z., & Busemeyer, J.R. (2013). A quantum question order model supported by empirical tests of an a priori and precise prediction. *Topics in Cognitive Science*, Vol. 5, No. 4, pp. 689–710; DOI: 10.1111/tops.12040.

Wang, Z., & Busemeyer, J.R. (2015). Comparing quantum versus Markov random walk models of judgments measured by rating scales. *Philosophical Transactions of the Royal Society A: Mathematical, Physical and Engineering Sciences*, Vol. 374, No. 20150098; DOI: 10.1098/rsta.2015.0098.

White, L.C., Barque – Duran, A., & Pothos, E.M. (2016). An investigation of a quantum probability model for the constructive effect of affective evaluation. *Philosophical Transactions A*, Vol. 374, No. 20150142; DOI: 10.1098/rsta.2015.0142.

White, L., Pothos, E.M., & Busemeyer, J.R. (2014). Sometimes it does hurt to ask: The constructive role of articulating impressions. *Cognition*, Vol. 133, pp. 48–64; DOI: 10.1016/j.cognition.2014.05.015.

White, L.C., Pothos, E.M., & Jarrett, M. (2020). The cost of asking: How evaluations bias subsequent judgments. *Decision*, Vol. 7, No. 4, pp. 259–286; DOI: 10.1037/dec0000136.

Willen, R.M., & Stroemwall, L.A. (2012). Offenders' uncoerced false confessions: A new application of statement analysis? *Legal and Criminological Psychology*, Vol. 17, pp. 346–359; DOI: 10.1111/j.2044-8333.2011.02018.x.

Wojciechowski, B.W. (2012). *Ocena Wiarygodności Zeznań Świadków Dorosłych Przy Zastosowaniu Psychologicznych I Prawniczych Kryteriów Treściowych* [Veracity assessment of adult witnesses' statements with the use of psychological and legal content criteria], unpublished PhD thesis, The University of Silesia, Katowice.

Wojciechowski, B.W. (2014). Content analysis algorithms: An innovative and accurate approach to statement veracity assessment. *European Polygraph*, Vol. 8, pp. 119–128; DOI: 10.2478/ep-2014-0010.

Wojciechowski, B.W. (2016). *Analiza i Ocena Zeznań Świadków* [Analysis and assessment of witnesses' statements]. Gdańskie Wydawnictwo Psychologiczne, Sopot.

Wojciechowski, B.W., Gräns, M., &Lidén M. (2018) A true denial or a false confession? Assessing veracity of suspects' statements using MASAM and SVA. *PLOS ONE*, Vol. 13, No. 6, e0198211; DOI: 10.1371/journal.pone.0198211

Wojciechowski, B.W., Izydoryczk, B., Blasiak, P., Yearsley, J., White, L.C., & Pothos, E.M. (2022). Constructive biases in clinical judgment. *Topics in Cognitive Science*, Vol. 14, pp. 508–527; DOI: 10.1111/tops.12547.

7

SUSPECTS' STRATEGIES AND BELL INEQUALITIES

Suspects' trial strategies

One of the fundamental principles of the criminal process is the presumption of innocence, which requires that during criminal proceedings it has to be determined whether there are sufficient grounds to hold a particular individual criminally liable. Thus, it is the responsibility of law enforcement to gather evidence against the suspect (accused) and establish guilt. From this understanding of the criminal process and the presumption of innocence derive certain rights of a charged individual, including the right to actively defend oneself and fight the charges against oneself. Among the most important rights of a suspect under the provisions of the Code of Criminal Procedure is the right to refuse to give explanations and answer questions. According to this principle, concealing information or providing a false account of events cannot result in a harsher sentence. According to the principles, the suspect is not required to provide evidence against himself or to demonstrate his innocence. In addition, a person who has violated the law is not required to tell the truth and is not subject to criminal liability for giving false testimony, as stipulated in the Criminal Code.

The suspect's strategy and the procedural situation may have a significant impact on the explanations provided and any potential alterations. The suspect may give or alter an explanation based on the gravity of the crime, the likelihood of conviction, the criminal sanctions that may be imposed or the social or even familial consequences of an admission of guilt, both in the short and long term. Given his or her circumstances, the suspect may seek to end the proceedings as quickly as possible or to delay them; seek the possibility of legal leniency or measures such as conditional dismissal or conviction without trial (Wojciechowski, 2015).

DOI: 10.4324/9781003402480-7

Current research on the strategies of criminal suspects indicates that previously convicted individuals (who have served a prison sentence) have different procedural strategies than those who have never been convicted. One of the earliest studies on this topic revealed that previously convicted offenders were more likely to exercise their right to remain silent during interrogation (Softley, 1980). Another study conducted by Leo (1996) showed that offenders without a criminal record or previously punished for minor offences were more likely to waive the suspects' rights to legal counsel or their right to remain silent (Miranda rights). This study has additional value because it was conducted in the field using a participatory observation method. Leo analysed 153 interviews regarding fraud, burglary and theft, in addition to domestic violence and assault. He concluded that a suspect is more likely to exercise his rights the more familiar he is with the legal system (refusal to explain and assistance of a lawyer). Strömwall and Willén (2011) investigated the trial strategies employed by convicted individuals. The authors conducted interviews with prison inmates and discovered that one of the primary strategies was not to divulge any information to the law enforcement officers. Such a strategy is highly adaptable and useful, as it avoids providing a self-incriminating explanation while allowing the prisoner to view the evidence in police possession. The second reported major strategy was to stay close to the truth. The findings of this study indicate that previously convicted individuals are aware of how the information they provide during an interview affects the consideration of their case, the likelihood of conviction and the final determination of guilt. This also demonstrates that they have a solid grasp of how the justice system operates and the rules that govern the criminal justice process. This hypothesis is also supported by research conducted by Granhag and colleagues (2009), which demonstrates that people without a criminal history provide more self-incriminating information than those with a criminal history. In addition, the tendency for first-time interviewees to provide self-incriminating explanations against the suspect persisted regardless of the evidence available to the police and the manner in which the interview was conducted.

Rohrbaugh and colleagues (1999) proposed a model to describe the relationship between various factors determining suspects' strategies and selection processes. The model was created to explain why subjects make different decisions when confronted with the same problem under identical conditions. According to Rogrbaugh and colleagues (1999), differences in decision-making are the result of the interaction of three factors: context (environmental factors external to the decision-maker that influence the course of cognitive processes), cognitive processes (information processing) and individual variables, particularly prior experience. The researchers hypothesize that the information deemed diagnostic or crucial to a decision and leading to the conclusion of the decision-making process depends on the interaction of the aforementioned three factors. In addition, there is research on the influence of the emotional context surrounding the decision-making process on choices. Feelings of guilt or shame in respondents have a direct effect on their intentions, it can also manifest as a postponement of the moment of decision-making, such as when the decision-maker is angry or anxious.

According to Peters and colleagues (2006), emotions serve four purposes during the decision-making process: as a source of information, motivation, comparison (common currency), and attentional control. The assumption that emotions act as a mediating variable in the decision-making process, which is associated with the emotional framing effect, is not debatable. Researchers have demonstrated that media reports, among other things, can have this effect and influence political opinions, attitudes and behaviour (Gross, 2008; Holm, 2012; Lecheler et al., 2015). It is important to note that this effect can be observed with the use of functional magnetic resonance imaging (Li et al., 2017).

Research indicates that decision-making processes are also highly context - dependent. Study on expected utility theory has shown, among other things, that contextual factors help to explain Allais paradox (Kahneman, 2011). One of the central assumptions of prospect theory is also that context influences decisions. The decision-makers resources, such as life or financial conditions, form a specific context for a decision; depending on the state of one's resources, a particular offer by another person can be viewed as either a gain or a loss. Image Theory by Beach (1993) was founded on this premise. According to this concept, each decision-maker possesses three information constructs known as images. These pertain to moral rules, the goal that a given action is intended to bring the decision-maker closer to, and the action plan, i.e., the concrete steps that must be taken to achieve the desired effect in accordance with the moral rules.

Suspects' trial strategies were also analysed in the light of the game theory. A central aspect of the game theory is the problem of decisions made in multi-player settings, where each player has his or her own preferences (rules) that govern his or her behaviour. In turn, a participant's behaviour impacts his or her payoff. The payoff for a player can be an advantage in the form of a reward or the avoidance of a penalty. According to economic theory, each player in a game acts rationally and makes decisions that maximize his or her own benefit, regardless of the moves of the other players. Nevertheless, from the perspective of the social sciences, game theory is a significantly more complex phenomenon than a simple mathematical equation. Numerous scientific disciplines, including sociology, biology, political science and computer science, can utilize it. In all of these scientific fields, game theory is used to examine various models relating to optimal decision-making; however, the determining factors in each field are entirely unique and dependent on numerous variables. The foundation of game theory is uncertainty regarding the actions of other players. Player A cannot be convinced that Player B has not also maximized his winnings by reducing Player A's winnings in order to maximize his own. We "play games" almost constantly in daily life. Sometimes these are two-player games, frequently they are multi-player games, and sometimes they are even single-player games in which only one player acts under uncertain conditions and makes decisions that affect his payout and the payouts of the other players.

The most well-known example is the prisoner's dilemma (Tucker, 1950). It is based on a situation in which two criminal accomplices are interrogated simultaneously. Each may either admit guilt while framing the other or remain silent. If only one of them confessed, he would be released and the other would spend six years in prison. Alternatively, if neither turns the other in, they will both spend a year in prison. In the event that both prisoners confess, they will each be sentenced to three years in prison. A typically symmetrical payoff matrix determines each player's reward based on their collective action. Typically, payoffs are designed so that it is most advantageous for defect if the other player chooses cooperation, but mutual gain is greatest if both players choose cooperation defection is then the Nash equilibrium. The optimal strategy for both parties is to remain silent, provided that they have complete confidence in one another and that neither will confess (Chater et al., 2008). Prisoner's dilemma games have been studied extensively in psychology, in part because they can lead to apparent discrepancies with classical probability theory, violating the law of total probability (Broekaert et al., 2020; Chater et al., 2008; Griffiths et al., 2010; Khrennikov, 2004; Oaksford & Chater, 2009; Pothos & Busemeyer, 2009; Shafir & Tversky, 1992; Tenenbaum et al., 2011).

Entanglement, supercorrelations and Bell inequalities

Quantum cognition is a framework for building cognitive models based on the mathematics of quantum probability theory, which is, like classical probability theory, a mathematical framework for assigning probabilities to events (Busemeyer & Bruza, 2012). In its most basic form, quantum cognition is merely an algorithm for calculating probabilities, with no claim to replicate the brain's underlying decision-making process. According to this perspective, the success or failure of a method is determined solely by the empirical accuracy of its forecasts. Quantum cognition comprises knowledge about the set of conceivable dynamics, state changes and measurements that can be conducted on a system, but this comes directly from the fundamental probabilistic structure to a great extent (Yearsley & Busemeyer, 2016).

At the heart of quantum theory is a phenomenon called entanglement. Quantum entanglement refers to a specific sort of quantum superposition. Entangled states exhibit a causal nonlocal correlations between the outcomes of measurements on spatially distinct subsystems of a system. It is a relation in which one of the elements of the system (i.e., entangled particle) depends on the aspects of the other entangled element. When two particles, such as a pair of photons or electrons, become entangled, they maintain their connection despite being separated by great distances. The relationship between particles generates entanglement. Scientists refer to this as an emergent feature. Prior to being measured, each particle can be in a state of superposition, or both "spin up" and "spin down" simultaneously. If the researcher detects the direction of one particle's spin and then repeats the measurement on the particle's distant, entangled partner, the pair will always be correlated: if one particle's spin is up, the other will be down (note, that the spins may instead both be up or both be down and at different angles, depending on how the experiment is designed, but there will always be a correlation).

Entangled systems can result from any interaction, and entanglement is a universal property of the quantum world. Quantum entanglement is pervasive and fundamental to quantum theory. Due to the ubiquitous nature of these correlations in entangled systems, such systems are also referred to be non-separable. Nonlocal correlations are the foundation of quantum holism, often known as nonlocal correlations. It indicates that the primary difficulty in comprehending quantum systems is how to decompose them into subsystems, not how to assemble them from simple components. Decomposing an entangled system yields subsystems that are "nonlocally" coupled (i.e., without direct causal interactions) (Wang et al., 2013).

Einstein, Podolsky and Rosen presented a thought experiment in a paper (1935) to suggest that quantum mechanics was not a comprehensive physical theory. The thought experiment, known now as the "EPR paradox," was designed to illustrate the inherent conceptual challenges of quantum theory. It was said that the result of a measurement on one particle of an entangled quantum system can have an instantaneous effect on another particle, despite the distance between them. One of the defining characteristics of quantum physics, beside entanglement, is the concept of uncertainty: not all the classical physical observable qualities of a system can be simultaneously and precisely determined, even in principle. As an alternative, there may be multiple sets of observable attributes, such as position and momentum, that cannot be known simultaneously. If, for instance, two photons get entangled – that is, if they are permitted to interact initially in order to be described by a single wave function – then, even after they are separated, they will continue to share a wave function. So, measuring one will determine the state of the other; for instance, in a spin-zero entangled state, if one particle is measured to be in a spin-up state, the other is immediately compelled to be in a spin-down state. This is known as the nonlocal behaviour. Because it appears to violate one of the central tenets of relativity theory – that nothing, even information cannot be transformed faster than the speed of light, because this would violate causality – Einstein called it a "spooky action at a distance."

EPR paradox is a result of a thought experiment designed by the authors to show that quantum mechanics could not be considered a complete theory of nature (Einstein et al., 1935). In Einstein's and colleagues' opinion (1935;

Bell, 1964, 1987), a hidden variable can explain the behaviour of the entangled particles, an existence of an unknown local property of the system which should account for the discrepancy would make the "spooky action" unnecessary. The thought experiment contested the quantum–mechanical notion that simultaneous values of incompatible observables are not only impossible to know, but also pointless to ponder. Niels Bohr, who made foundational contributions to understanding the quantum theory disagreed with Einstein and defended a strict Copenhagen interpretation of quantum mechanics, rejecting the assumption that hidden variables are necessary to explain results of the experiments on entangled systems.

Einstein, Podolsky and Rosen's objections to the actuality of unmeasured characteristics were purely philosophical for nearly three decades, but contributed considerably to the development of quantum theory and knowledge of quantum mechanics (Mermin, 1981). EPR paradox exposed one of the fundamental non-classical characteristics of the measurement process – measurement can be performed on a distant entangled particle and therefore without disturbing directly the second entangled particle. John Bell (1964, 1987) showed that this dispute can be empirically resolved by measuring the size of the correlations between pairs of variables in quantum systems that are entangled. Bell's inequalities specify a limit for classical correlations that is projected to be violated by quantum systems that are entangled. Currently, these nonlocal quantum correlations are empirically well-established in physics, and novel applications of them have been discovered (Aspect et al., 1982; Gisin, 2021; Guehne et al., 2010; Gustafson, 2005; Saito, 2004; Vedral, 2006). Typically, Bell inequalities are formulated using measurements of pairs of spatially separated variables. However, it is also possible to develop temporal Bell inequalities for a single variable separated in time, for example at different point on timeline (Leggett & Garg, 1985). Violations of temporal Bell inequalities would demonstrate that system states cannot be precisely confined in time, as classical theory would always require.

Bell inequalities are usually explained in literature by looking at a system made up of two smaller systems. A pair of people, like Alice and Bob, can be thought of as a composite system, with each person as a subsystem. These two people can talk to each other for a while and then go on their separate ways. For instance, the two people could be two prisoners who want to escape from jail. They talk to each other as they make their plan, and then they go their own ways to carry out the plan. Alice and Bob are the names of the two prisoners. Think of two sets of two (binary) questions we could ask Alice and Bob, with A and A' for Alice and B and B' for Bob. There can be a lot of overlap between the questions. For example, let's say that both questions are about the escape plan and whether or not Alice and Bob are likely to be safe if they go their own ways. We know that Alice and Bob talked to each other before they broke up. So, we already think that their answers will be co-dependent, that is, that the answers for A, A', and B, B' will be similar to each other. In terms of physics, and assuming a classical picture, Alice and Bob's answers have a common cause, which can result in correlations between their answers. Bell's (1964) very clever and important analysis of math shows that there is a limit because of how A, A' and B, B' are related to each other. So long as the correlations between A, A', B, and B' are caused by (just) one thing in common, the size of these correlations must be limited to 2. Equations for different variants of the physical experiment can be written in the following inequalities (Bell, 1964; Blasiak et al., 2021; Clauser et al., 1969):

$$S_1 = \langle A, B \rangle_{00} + \langle A, B' \rangle_{01} + \langle A', B \rangle_{10} - \langle A', B' \rangle_{11}$$
$$S_2 = \langle A, B \rangle_{00} + \langle A, B' \rangle_{01} - \langle A', B \rangle_{10} + \langle A', B' \rangle_{11}$$
$$S_3 = \langle A, B \rangle_{00} - \langle A, B' \rangle_{01} + \langle A', B \rangle_{10} + \langle A', B' \rangle_{11}$$
$$S_4 = - \langle A, B \rangle_{00} + \langle A, B' \rangle_{01} + \langle A', B \rangle_{10} + \langle A', B' \rangle_{11}$$

For each equation, limit set by the Bell test is equal to two ($S \leq 2$) (Clauser et al., 1969; Fine, 1982). Even if Alice and Bob before the interrogation have developed a response strategy, level of conformity of their answers, according to Bell, cannot exceed the set limit of two. As long as there is one common cause of the correlations between Alice's answers to question 1 or 2 (A, A') and Bob's answers to question 1 or 2 (B, B'), no matter how many questions they will be asked, what will be the exact set of questions (i.e., A&B' or A'&B') and how many times the interrogation will be repeated, the size of correlations and level of concordance cannot exceed limit of 2. Any interaction that would violate the Bell bounds, that is correlations that exceed the limit of two, would indicate that the state of the composite system has to be entangled.

In physics, the derivation of the inequalities in a Bell experiment is based on the assumptions of realism, locality and free choice; any observed violations demonstrate that models that insist on all three assumptions contradict physical reality (Gisin, 2021; Popescu & Rohrlich, 1994). Locality and freedom of choice are characteristics of the two statistically important systems. The principle of locality states that an object is influenced only by its immediate surroundings and if two objects have effect on each other, the space between those objects must mediate the action. Hence, in the example with Alice and Bob, locality indicates that the two agents are local relative to one another – there is no communication – so that Alice has no information about Bob when making her decisions and vice versa. Free choice is connected with the setup of an experiment – experimenter has to be able to choose freely which component of the particle's spin (at the *x*, *y* or *z* axis) should be measured. Free choice implies that Alice's decisions are not affected by Bob's (Waddup et al., 2021). Bell experiments are usually conducted by measuring two spatially and temporally separated components of an entangled particle system (e.g. spin measurements). In all of the recent experiments conducted to test the Bell inequalities, it is necessary to assure that researcher's free choice. Observables or measurements cannot commute with each other, both quantities cannot be recorded at the same time without trade-off in the accuracy in the measurement, and as a result, it is not possible to acquire simultaneously precise information from them (Zhang et al., 2022). Because answering one question creates a unique context for answering the other, same questions could have different meaning when in insolation, versus when asked after another incompatible question was answered. In other words, answer to the first question alters the context for the other question and it is not possible to create a joint probability distribution for the system (Pothos & Busemeyer, 2013). Therefore, when Bell test is conducted, the measurements of the entangled particles must be incompatible, they have to be conducted independently, with the use of devices which enable measurement of one entangled particle without any effect on the other (Heinosaari et al., 2016). Realism relates to assumption that properties of objects are real and exist in our physical universe independent from observer's mind and independent from measurement.

Bell bounds can be broken when Alice and Bob are perfectly tuned to each other, so that when answering to questions one and two they give entangled answers: for questions A&B answer would be 1 if replies confirm entanglement, same for questions A&B' and A'&B. If locality and free choice apply, then set of questions A'&B' should be answered with correlation as well. This is result of perfect fit of the other three types of questions A&B, A&B' and A'&B. However, if answers to questions A'&B' are consistent, then the sum of points is set by the limit S=2, which is the maximum value that S can take, assuming realism, locality and free choice. Therefore, the only way the Bell's bound can be broken is by some kind of sensitivity to context. For example, Bob's answers are sensitive to the context created by Alice's questions (Blasiak et al., 2021; see also: Gallus et al., 2023; Waddup et al., 2021).

Blasiak and colleagues (2021) introduced sensitivity to context, supposing that the B' question answered by Bob depends on whether Alice considers question A or A'. If Alice answers question A then Bob responds to B' in a way that the two questions correlate with each other, (A&B') = 1. However, if Alice considers the A' question, then Bob responds to B' in a way that the outcomes of the two questions have to anticorrelate, and as a result in the case of this scenario A'&B'= -1. That is, there is no Bob's answer to the B' question, independently of what Alice does. If we accept the possibility of sensitivity to context, then we can easily see that the Bell bound can be exceeded, since:

if

$$S=\langle A, B\rangle_{00} + \langle A, B'\rangle_{01} + \langle A', B\rangle_{10} - \langle A', B'\rangle_{11}$$

then

$$S=1 + 1 + 1 - (-1) = 1+ 1 + 1 + 1 = 4$$

This type of correlation produces S = 4, and it is the strongest nonlocal correlation that is not a result of signalling. This type of nonlocal correlation in the two questions, two outcomes (answers) scenario in the physical literature it is called the PR box. It is the strongest type of nonlocal correlations known to science (Popescu & Rohrlich, 1994).

PR box can be also used to further describe and explain sensitivity to context and supercorrelations. Sensitivity to context means that the original set of questions A, A', B, B' is better understood when we show them as collection of following questions: A, A', B, B'&A, B'&A', where question B' has two different versions: first version when Alice responds to question A, and the second when Alice responds to question A'. For S > 2 it is not sufficient for pairs of questions to be responded perfectly in tune with each other (this would be a case of perfect, classical correlation). It is also required that responses are sensitive to the questions asked by the other agent. Pothos and colleagues (in: Blasiak et al., 2021; Gallus et al., 2023; Waddup et al., 2021) proposed to treat cases of sum of answers exceeding limit set by Bell (S> 2) as reflecting the supercorrelations. The difference between correlations know in different scientific disciplines and the supercorrelations studied in the Bell tests context is that classical correlation refers to binary relation, whereas supercorrelation is a relation between answers amongst two sets of questions. In classical correlations, we are analysing only relation between questions A and B, and in the supercorrelations experiments participants answer two types of questions A/B and A'/B'. We can conclude that a supercorrelation occurred, when co-dependence is observed between answers given by Alice and Bob even, when the answer different questions.

There are two types of interactions that can result in supercorrelations above the Bell constraints. The first lacks both physical and psychological appeal. Assume that Alice and Bob are directly influenced by one another. A direct influence involves the physical transmission of information from one person to another. Hence, while Alice is answering the A, A' questions, Bob may be answering the B, B' questions on the phone. In such a straightforward scenario, correlations between Alice's and Bob's responses that exceed what would be predicted based on a common cause are scarcely surprising. In physical systems, a direct impact may equate to a tangible (though sometimes concealed) connection between subsystems (e.g., Aerts, 1982).

Suppose, however, that Alice and Bob are in fact completely separated (maybe by light years) and are unable to interact with one another at the time of issue resolution. If Alice and Bob were microphysical systems regulated by quantum mechanics, the theory of quantum mechanics predicts, under certain conditions, that the correlations between A, A', B and B' can exceed the Bell constraint; this prediction has been empirically verified (Aspect et al., 1982). To further clarify, we can examine a situation involving two people who had only communicated previous to their separation. Bell's theorem states that, classically, correlations between A, A', B and B' must be compatible with Bell's bound (S≤2) so long as the influence between Alice and Bob involves solely a shared source prior to separation (and we can exclude any direct effect between Alistair and Bob). Due to the proof that they are not, physicists have labelled any future interaction between Alice and Bob as a supernatural action at a distance. Entangled states and the violation of the Bell bound on correlations give rise to so called "spooky action at the distance." These correlations are referred regarded as supercorrelations to indicate that they cannot emerge in a typical manner. By answering their queries, Alice and Bob appear to be communicating despite the fact that there is no physical means by which this might be accomplished (Blasiak et al., 2021). It is worth to mention that though entangled particles can violate Bell's inequality, there is a weaker inequality of similar type setting the upper bound on supercorrelation. It is called the Tsirelson's bound, and it is equal to around 2.828, that is the product of two times square root of two (S≤2*√2) (Tsirelson, 1980).

Alain Aspect, John Clauser and Anthon Zeilinger have conducted physical experiments to establish, if Bell's inequalities can be violated. In a series of studies on entangled photos, they have confirmed the fundamental prediction of quantum mechanics and provided evidence that in quantum system it is possible to break the borders set by the Bell inequality; importance of their research has been recognized by the 2022 Nobel Prize in Physics. Bell test is a quantum – world physics experiment designed to test theory of quantum mechanics, but Aerts (1982) gave an example of a classical macroscopical situation that can violate Bell inequalities. Analysis conducted by Gallus and colleagues (2023) demonstrated not only, that Bell – type tests are applicable as general statistical technique, without any domain specificity, but also showed evidence that boundaries set by the Bell test are being crossed in the macro scale.

There have been also tests of these concepts in psychology (Khrennikov, 2022; Waddup et al., 2021). Aerts and Sozzo (2011) conducted a study in the area of conceptual combination to verify if pairs of concepts can display patterns of correlations that are stronger than what is allowed classically. Cervantes and Dzhafarov (2017) conducted an experiment in which respondents were making choices between characters of a story ("*The Snow Queen*" by H.Ch. Andersen) and matching them characteristics (such as: kind or evil). Results show that quantum-like probabilistic contextuality can be applied in psychology, choices that participants made were directly influencing each other. In various experiments based on the design introduced by Cervantes and Dzhafarov, contextuality in simple decision-making was observed (Basieva et al., 2019). Cognitive experiments by Aerts and colleagues (2016) confirmed deviations (including conceptual vagueness, underextension and overextension) from the rules of classical logic and probability theory. Yearsley and Pothos (2014) explored use of temporal Bell (Leggett–Garg) inequalities in cognition, modelling changes of the mental representations with time. Study conducted by Aerts and Sozzo (2018) in which participants were asked to pick a pair of wind directions, considered by them as an example of two different wind directions, showed violations of the Bell inequalities.

In one of the most recent papers, Waddup and colleagues (2021) for the first time addressed whether entanglement and Bell inequalities can be applied in the case of the interaction of distinct agents, physically isolated from each other. After conducting a series of experiments with a variants of prisoner's dilemma tasks to test sensitivity of participants to context, the authors concluded that: observed behaviour appears inconsistent with baseline classical probability theory, sensitivity to context was observed and study participants demonstrated behaviour similar to entangled particles, Bell bound was exceeded and quantum model fitted superior to the collected data.

Study on suspects' strategies

To test if offenders are sensitive to context and if supercorrelations breaking the Bell bounds can be observed in legal settings, we have conducted study on two groups of perpetrators – acting alone and committing crimes jointly and in concert with others. Our research paradigm was based on design proposed by Waddup and colleagues (2021), since it was the first research exploring Bell's ideas for interacting agents, and the theoretical basis was the model by Blasiak et al. (2021), developed for the analysis and modelling of sensitivity to context.

We will view pairs of suspects as a composite system, with each individual functioning as a subsystem. We will focus on participants which have previously interacted, before being separated – committed together a crime and compare pairs of offenders which previously committed a crime, but without the involvement of others. We can anticipate that there will be some level of correlation in responses to questions, but interesting question is if the level of concordance will fit within borders of the classical picture or as a result of previous interaction offenders can show supercorrelation, breaking Bell bounds.

How can Bell's bound be broken? For macroscopic, interacting agents, like offender A and offender B, the only way Bell's bound can be exceeded is if at least one of the free choice or locality assumptions is violated. In practice, for $S>2$, the second offender (B) needs to adjust his answers depending, first of all on which question the first offender (A) answered (assuming free choice, i.e., that there is no correlation between the questions that B and A receive) and provide correct answer. That is, offender B's answers have to be contextual, relative to offender A's questions. When suspect A considers the first question, offender B should respond to the second question in a way that the two questions correlate with each other. However, when offender A considers the second question, then offender B has to respond to the second question in such a way that the outcomes of the two answers anticorrelate (A'&B') = −1. That is, there is no answer to the B' question, independent of what the first suspect does. If we accept the possibility of this contextuality, then we can easily see that theoretically the Bell bound can be exceeded, in that:

$$S = \langle A, B\rangle_{00} + \langle A, B'\rangle_{01} + \langle A', B\rangle_{10} - \langle A', B'\rangle_{11} = 1 + 1 + 1 - (-1)$$
$$= 1 + 1 + 1 + 1 = 4$$

It is generally agreed, that contextuality is impossible for macroscopic systems, if locality and/or free choice are enforced. However, without one of these assumptions, offender B can have knowledge of offender's A questions, and so adjust his strategy accordingly. For example, B might be able to spy on A (violating locality) to determine his question and given example.

Material and methods

The basic paradigm of the experiment was similar to Waddup et al. (2021), with few modifications. We were willing to conduct a study on perpetrators of crimes, and compare answers given by pairs of knowing each other and anonymous participants. We have also made an attempt to prepare set of incompatible questions; although we are aware, that it is debatable whether we have achieved this goal. And last but not least, we did not allow participants to check answers given by their counterparts and tried to eliminate risk of signalling.

There were two individuals answering independently binary questions, such that different combinations of answers would result in different payoffs. A variation of the prisoner's dilemma was employed. Both participants read summary of a criminal records describing robbery and were instructed that they are suspected of committing this crime with the second offender. They were also informed that they were arrested and will be interviewed by the prosecutor. Participants were warned that their associates will be interviewed separately and questions regarding three subjects will be asked: whether they were present at the crime scene (A_1&B_1) and whether their image could have been captured on a surveillance camera installed at the road intersection where the offence occurred (A_1'&B_1'); when was the last time, they met the other suspect (A_2&B_2) and what meal they had together (A_2'&B_2'); in which direction they have travelled on the night of the crime (A_3&B_3) and what means of transportation they have used on the night of the event (A_3'&B_3').

There were multiple types of trials. Sometimes, participants and their associates received the same questions, at other times different questions. Questions for different pairs of participants were asked in different order and different variations. For example, one pair of participants would be asked A_2, A_3, A_1' and B_1', B_2,B_3' set of questions, meaning that the first participant would be asked about the last meeting with the other suspect (A_2), direction of travel on the night of the event (A_3) and whether he was at the crime scene (A_1'), and the second suspect would be asked about whether his image was recorded by the surveillance camera (B_1'), when was the last time he met accomplice (B_2) and what means of transpiration he used on the night of the event (B_3'). The instructions indicated that the perpetrators of the robbery were present at the scene, and their image may have been captured by surveillance cameras. In addition, law enforcement knew from secured traces that the perpetrators of the crime last met on Monday and had breakfast together. It was also established by the police that offenders after committing the crime travelled north by car.

According to the instruction and case summary given to the participants, answers could indicate their relation to the other suspect and guilt (confess) or constitute an attempt to dismiss the charges and undermine the existence of any links to the other suspect (deny). Participants of the study were also instructed that depending on the combination of questions, their answers and their associates' questions and responses, different sentencing policy would apply. We assumed that participants would favour decisions leading to lower sentences (for themselves or for both themselves and their associate). From the wording of the instructions, it was clear that to file charges, it was sufficient for law enforcement agencies to find links to crime in response to even just one question and even one answer to a question is enough to attribute to them a connection to the case. Participants were also instructed about penalty. Standard prisoner's dilemma payoff matrix was applied (Poundstone, 1993), leading to a possible of four different outcomes:

- if A and B both deny, they will each serve the lesser charge of two years in prison;
- if A confesses B but B denies, A will be set free while B serves ten years in prison;
- if A denies but B confesses A, A will serve ten years in prison and B will be set free;
- if A and B both confess, they share the sentence and serve five years.

Table 7.1. shows an example of a questions set A_1, A_2', A_3' and B_1', B_2,B_3' referring to the three aspects of crime and how the sentencing policy would vary, given a specific set of answers for suspect A and suspect B.

For the experiment, persons serving prison sentences were recruited, with help of attorneys at law (defence counsels) personally known to the researcher. Participation in the research was optional. Participants were informed that the purpose of the study was to analyse suspect's strategies. Prior to the study, participants received a consent form and, following its conclusion, a debriefing form. The participants were not compensated for their participation. As it was impossible to pay the participants, we determined that none of them would be

TABLE 7.1 Example of set of questions and sentencing policies which would apply, depending on the various answers

	Suspect A		*Suspect B*		*Suspect A's sentence*	*Suspect B's sentence*
	Question	*Answer*	*Question*	*Answer*		
A_1&B_1'	Were you present at the crime scene: *Yes or No*	Yes	Is it possible that your image has been captured by surveillance camera at the crime scene *Yes or No*	No	0 years	Ten years
A_2'&B_2	What was the last meal you had together with the other suspect: *Breakfast (on Monday) or Dinner (on Tuesday)*	Dinner	When was the last time you met the other suspect: *Monday (at breakfast) or Tuesday (at dinner))*	Monday	Ten years	0 years
A_3'&B_3'	What means of transportation have you used on the night of the event: Bus or Car	Car	What means of transportation have you used on the night of the event: Bus or Car	Car	Two years	Two years

compensated. Upon request participants could receive a book or a magazine, unavailable in the prison's library.

Participants received paper copies of research material from their defence counsel during a meeting, that is: consent form, instruction which explained the context, set of simple demographics questions and one round of questions. Subjects were assured that participation in the experiment is voluntary, and were informed about purpose of the study, steps of the trial, meaning of the answers and payoff matrix. Answers given by participants included in the first group were paired with co-offenders (persons with which they committed a crime in the past). In the second group subjects were randomly assigned to the pairs of participants serving time for individually committed crimes. The instruction for those answering in pairs with known co-conspirators indicated that the answer choices made could affect not only the legal situation of the person answering, but also that of the co-conspirator.

All tasks have been completed during the meeting with the defence attorney. In conditions that allow questions to be answered freely, without the participation of third parties. After solving the tasks, participants packed the answer sheets into an envelope and handed them closed in the sealed envelope to their counsellor to ensure anonymity.

Participants

Two groups of male offenders participated in the study: those serving time for crimes done in collaboration with others and those previously penalized purely for solo offences. The first group consisted of 64 men between the ages of 23 and 61 (M= 36, SD = 9.87). The second group comprised 66 men between the ages of 22 and 59 (M=35, SD=13.47). The study group was dominated by those with vocational education (48.9%), followed by those with primary education (27.8%); those with secondary education (18.2%) and those with higher education (5.1%) made up the sample, respectively.

Results

The experiment was focused on obtaining statistics on decision choices which reflect contextuality. To achieve that, we have asked participants to answer questions referring to three aspects of the crime, however for each theme there were two possible options of questions. To verify if previous interaction can have effect on the results, we have collected answers from 64 paired offenders having previous criminal history (previously sentenced for cooperation), and 66 participants serving a prison sentence, unknown to each other. Overall, there were 192 answers from the familiar pairs of suspects and 198 answers from strangers.

Table 7.2 presents distribution of answers to questions regarding three aspects of the crime, given by research participants between whom there was a pre-existing relationship and strangers. For each pair of matching answers to pairs of questions (A&B, A'&B, A&B' or A'&B') which would allow respondents to protect themselves from criminal liability – one point was granted; for answers that could lead to a conviction – zero points were granted. Values presented in the Table 7.2 represent the overall observed concordance of responses – ratio of the number of times when participant denied together with his counterpart to the overall number of answers. A result equal to.50 would indicate that in 50% of cases answers given by paired offenders would allow them to avoid criminal liability and in 50% of cases their answers would confirm their connection to the crime.

We have computed the S values separately for each of the groups and each of the pair of questions (themes). For the first pair of questions ("Were you present at the crime scene" and "Is it possible that your image has been captured by surveillance camera at the crime scene"), the S value for offenders with a shared criminal history was 1.81, and for stranger perpetrators, it was 1.77; the difference between the groups was insignificant ($t(128)=1.06$; $p=.29$). For the second pair of questions ("When was the last time you met the other suspect, Monday (at breakfast) or Tuesday (at dinner)" and "What was the last meal you had together with the other suspect, breakfast (on Monday) or dinner (on Tuesday)") the S value for the participants knowing each

TABLE 7.2 Observed results for two groups of pairs of participants

	Offenders with a shared criminal history (N=64)			*Strangers (N=66)*		
Theme 1	*A/B: Were you present at the crime scene,* Yes or No *A'/B': Is it possible that your image has been captured by surveillance camera at the crime scene, Yes or No*			*A/B: Were you present at the crime scene, Yes or No* *A'/B': Is it possible that your image has been captured by surveillance camera at the crime scene, Yes or No*		
		B	B'		B	B'
	A	.46	.43	A	.47	.42
	A'	.48	.44	A'	.44	.44
Theme 2	*A/B: When was the last time you met the other suspect, Monday (at breakfast) or Tuesday (at dinner))* *A'/B': What was the last meal you had together with the other suspect, breakfast (on Monday) or dinner (on Tuesday)*			*A/B: When was the last time you met the other suspect, Monday (at breakfast) or Tuesday (at dinner))* *A'/B': What was the last meal you had together with the other suspect, breakfast (on Monday) or dinner (on Tuesday)*		
		B	B'		B	B'
	A	.44	.42	A	.39	.41
	A'	.43	.39	A'	.43	.44
Theme 3	*A/B: What means of transportation have you used on the night of the event, bus or car* *A'&B': In what direction did you travel on the night of the incident, North or South*			*A/B: What means of transportation have you used on the night of the event, bus or car* *A'&B': In what direction did you travel on the night of the incident, North or South*		
		B	B'		B	B'
	A	.45	.43	A	.45	.41
	A'	.42	.44	A'	.41	.43

other was 1.68, and for strangers, it was 1.67; the difference was insignificant ($t(128)=.26$, $p=.79$). Insignificant differences between the two compared groups were observed also for the third set of questions (about means of transportation and escape direction), the S value for affiliates was 1.74 and for strangers 1.70 ($t(128)=1.06$; $p=.29$).

To test if sensitivity to context can be observed in the participants' answers, we have compared differences between the answers to the B' questions depending on whether the first offender answered question A or A'. We could expect that answers to the B/B' questions will be different if the other suspect answered A or A' question. As a result of comparison of all three themes, we did not observe any significant differences between B' answers, linked to the first offender's A or A' answers.

The ability of a quantum system to break the classical S≤2 restriction (Bell bound), presumably without violating either locality or free will, is one of the most important discoveries in the history of quantum theory. This is confusing since to break the limit one system would have to respond to measurements in a way that is influenced by the measurements of the other system. How would this be possible for particles separated by vast distances? Can a macroscopic system break the theoretical limit?

The data produced by the experiment did not confirm quantum-like entanglement between the participants, and there were no significant differences between offenders with a shared criminal history compared to strangers. Following Waddup et al.'s (2021) study, we could expect that Bell's bound would be broken if suspects' answers would be in a perfect, stronger than classical correlation, called the supercorrelation. If participants would be able to adjust their answers both to the questions asked to their counterparts, and to replies given by their partners. However, unlike in the previous experiment, we did not break the locality or free choice assumptions, by letting offenders check questions asked to their partners and given by them answers. Our goal was to establish, if we can observe sensitivity to context and what is the level of supercorrelation when previously convicted individuals try to build a defence strategy against allegations.

References

Aerts, D. (1982). Example of a macroscopical classical situation that violates Bell inequalities. *Lettere al Nuovo Cimento*, Vol. 34, pp. 107–111.

Aerts, D., Arguelles, J.A., Beltran, L., Geriente, S., Sassoli de Bianchi, M., Sozzo, S., & Veloz, T. (2018). Spin and Wind Directions I: Identifying entanglement in nature and cognition. *Foundations of Science*, Vol. 23, pp. 323–335; DOI: 10.1007/s10699-017-9528-9.

Aerts, D., & Sozzo, S. (2011). *Quantum Structure in Cognition: Why and How Concepts Are Entangled.* In Proceedings of the Quantum Interaction Conference. Springer.

Aerts, D., Sozzo, S., & Veloz, T. (2016). New fundamental evidence of non-classical structure in the combination of natural concepts. *Philosophical Transactions A*, Vol. 374, p. 20150095; DOI: 10.1098/rsta.2015.0095.

Aspect, A., Grangier, P., & Roger, G. (1982). Experimental realization of einstein-podolsky-rosen-bohm gedankenexperiment: A new violation of Bell's inequalities. *Physical Review Letters*, Vol. 49, pp. 91–94.

Basieva, I., Cervantes, V.H., Dzhafarov, E., & Khrennikov, A. (2019). True contextuality beats direct influences in human decision making. *Journal of Experimental Psychology General*, Vol. 148, No. 11, pp. 1925–1937; DOI: 10.1037/xge0000585

Beach, L.R. (1993). Image theory: An alternative to normative decision theory. *Advances in Consumer Research*, Vol. 20, pp. 235–238.

Bell, J.S. (1964). On the einstein-podolsky-rosen paradox. *Physics*, Vol. 1, pp. 195–200.

Bell, J.S. (1987). *Speakable and Unspeakable in Quantum Mechanics*. Cambridge University Press, Cambridge.

Blasiak, E.M., Pothos, J.M., Yearsley, C., Gallus, E., & Borsuk (2021). Violations of locality and free choice are equivalent resources in Bell experiments. *Proceedings of the National Academy of Sciences of the United States of America*, Vol. 118, No. 17, pp. e2020569118; DOI: 10.1073/pnas.2020569118.

Broekaert, J.B., Busemeyer, J.R., & Pothos, E.M. (2020). The disjunction effect in two-stage simulated gambles. An experimental study and comparison of a heuristic logistic, Markov and quantum-like model. *Cognitive Psychology*, Vol. 117, p. 101262; DOI: 10.1016/j.cogpsych.2019.101262.

Busemeyer, J.R., & Bruza, P. (2012). *Quantum Models of Cognition and Decision Making*. Cambridge University Press, Cambridge.

Cervantes, V.H., & Dzhafarov, E. (2017). Show queen is evil and beautiful: Experimental evidence for probabilistic contextuality in human choices. *Decision*, Vol. 5, pp. 193–204; DOI: 10.1037/dec0000095.

Chater, N., Vlaev, I., & Grinberg, M. (2008). A new consequence of Simpson's paradox: Stable cooperation in one-shot prisoner's dilemma from populations of individualistic learners. *Journal of Experimental Psychology: General*, Vol. 137, pp. 403–421; DOI: 10.1037/0096–3445.137.3.403.

Clauser, J.F., Horne, M.A., Shimony, A., & Holt, R.A. (1969). Proposed experiment to test local hidden-variable theories. *Physical Review Letters*, Vol. 23, pp. 880–884.

Einstein, A., Podolsky, B., & Rosen, N. (1935). Can quantum-mechanical description of physical reality be considered complete? *Physical Review*, Vol. 47, pp. 777–780.

Fine, A. (1982). Joint distributions, quantum correlations, and commuting observables. *Journal of Mathematical Physics*, Vol. 23, No. 7, pp. 1306–1319; DOI: 10. 1063/1.525514.

Gallus, C., Pothos, E.M., Blasiak, P., Yearsley, J.M., & Wojciechowski, B.W. (2023). Bell correlations outside physics. *Scientific Reports*, Vol. 13, No. 4394; DOI: 10. 1038/s41598-023-31441-x.

Gisin, N. (2021). Indeterminism in physics, classical chaos and bohmian mechanics: Are real numbers really real? *Erkenntnis*, Vol. 86, pp. 1469–1481; DOI: 10.1007/ s10670-019-00165-8.

Granhag, P.A., Clemens, F., & Strömwall, L.A. (2009a). The usual and the unusual suspects: Level of suspicion and counter-interrogation tactics. *Journal of Investigative Psychology and Offender Profiling*, Vol. 6, No. 2, pp. 129–137; DOI: 10.1002/ jip.101

Granhag, P.A., Clemens, F., & Strömwall, L.A. (2009b). The usual and the unusual suspects: Level of suspicion and counter-interrogation tactics. *Journal of Investigative Psychology and Offender Profiling*, Vol. 6, No. 2, pp. 129–137.

Griffiths, T.L., Chater, N., Kemp, C., Perfors, A., & Tenenbaum, J.B. (2010). Probabilistic models of cognition: exploring representations and inductive biases. *Trends in Cognitive Sciences*, Vol. 14, pp. 357–364; DOI: 10.1016/j.tics.2010.05.004.

Gross, K. (2008). Framing persuasive appeals: Episodic and thematic framing, emotional response, and policy opinion. *Political Psychology*, Vol. 29, No. 2, pp. 169–192.

Guehne, O., Kleinmann, M., Cabell, A., Larsson, J.A., Kirchmair, G., Zaehringer, F., Gerritsma, R., & Roos, Ch.F. (2010). Compatibility and noncontextuality for sequential measurements. *Physcial Review A*, Vol. 81, No. 022121; DOI: 10/1103/ PhysRevA.81/022121.

Gustafson, K. (2005). Bell and Zeno. *International Journal of Theoretical Physics*, Vol. 44, No. 11, pp. 1933–1942; DOI:10.1007/s10773-005-8971-5.

Heinosaari, T., Miyadera, T., & Ziman, M. (2016). An invitation to quantum incompatibility. *Journal of Physics A: Mathematical and Theoretical*, Vol. 49, No. 123001; DOI: 10.1088/1751-8113/49/12/123001.

Holm, E.M. (2012). *Emotions as Mediators of Framing Effects*. Politica, Aarhus.

Kahneman, D. (2011). *Thinking, Fast and Slow*. Penguin Books, London.

Khrennikov, A. (2004). On quantum-like probabilistic structure of mental information. *Open Systems and Information Dynamics*, Vol. 11, pp. 267–275; DOI:10.1023/b:opsy.0000047570.68941.9d.

Khrennikov, A. (2022). On applicability of quantum formalism to model decision making: Can cognitive signaling be compatible with quantum theory? *Entropy*, Vol. 24, p. 1592; DOI: 10.3390/e24111592.

Lecheler, S., Bos, L., & Vliegenthart, R. (2015). The mediating role of emotions: News framing effects on opinions about immigration. *Journalism & Mass Communication Quarterly*, Vol. 92, No. 4, pp. 812–838; DOI: 10.1177/1077699015596338.

Leggett, A.J., & Garg, A. (1985). Quantum mechanics versus macroscopic realism: Is the flux there when nobody looks? *Physical Review Letters*, Vol. 54, pp. 857–860; DOI:10.1103/PhysRevLett.54.857

Leo, R.A. (1996). Inside the interrogation room. *Journal of Criminal Law and Criminology*, Vol. 86, No. 2, pp. 266–303.

Li, R., Smith, D.V., Clithero, J.A., Venkatraman, V., Carter, R.M., & Huettel, S.A. (2017). Reason's enemy is not emotion: Engagement of cognitive control networks explains biases in gain/loss framing. *Journal of Neuroscience*, Vol. 37, No. 13, pp. 3588–3598; DOI: 10.1523/jneurosci.3486-16.2017.

Mermin. (1981). Bringing home the atomic world: Quantum mysteries for anybody. *American Journal of Physics*, Vol. 49, No. 10, pp. 940–943.

Oaksford, M., & Chater, N. (2009). Précis of Bayesian rationality: The probabilistic approach to human reasoning. *Behavioral and Brain Sciences*, Vol. 32, pp. 69–120; DOI: 10.1017/S0140525X09000284.

Peters, E., Vastfjall, D., Garling, T., & Slovic, P. (2006). Affect and decision making: A "hot" topic. *Journal of Behavioral Decision Making*, Vol. 19, pp. 79–85.

Popescu, S., & Rohrlich, D. (1994). Quantum nonlocality as an axiom. *Foundations of Physics*, Vol. 24, pp. 379–385.

Pothos, E.M., & Busemeyer, J.R. (2009). A quantum probability explanation for violations of 'rational' decision theory. *Proceedings of the Royal Society B*, Vol. 276, pp. 2171–2178; DOI: 10.1098/rspb.2009.0121.

Pothos, E.M., & Busemeyer, J.R. (2013). Can quantum probability provide a new direction for cognitive modeling? *Behavioral and Brain Sciences*, Vol. 36, pp. 255–327; DOI: 10.1017/S0140525X12001525.

Poundstone, W. (1993). *Prisoner's Dilemma*. Anchor, New York.

Rohrbaugh, C.C., Shanteau, J., & Hall, B. (1999). Context, process, and experience: Research on applied judgment and decision making, [in:] *Handbook of Applied Cognition*, (ed.) Durso, F., John Wiley, New York, pp. 115–139.

Saito, N. (2004). Theory of quantum measurement in terms of quantum chaos. *Journal of the Physical Society of Japan*, Vol. 73, No. 6, pp. 1476–1481; DOI: 10.1143/JPSJ.73.1476.

Shafir, E., & Tversky, A. (1992). Thinking through uncertainty: Nonconsequential reasoning and choice. *Cognitive Psychology*, Vol. 24, pp. 449–474.

Softley, P. (1980). Police interrogation: An observational study in four police stations, Home Office research studies, No. 61; HMSO, London.

Strömwall, L.A., & Willén, R.M. (2011). Inside criminal minds: Offenders' strategies when lying. *Journal of Investigative Psychology and Offender Profiling*, Vol. 8, No. 3, pp. 271–281; DOI: 10.1002/jip.148.

Tenenbaum, J.B, Kemp, C., Griffiths, T.L., & Goodman, N. (2011). How to grow a mind: Statistics, structure, and abstraction. *Science*, Vol. 331, pp. 1279–1285; DOI:10.1126/science.1192788.

Tsirelson, B.S. (1980). Quantum generalizations of Bell's inequality. *Letters in Mathematical Physics*, Vol. 4, pp. 93–100.

Tucker, W.A. (1950), *Prisoner's Dilemma*, University of St. Thomas, Houston.

Vedral,V. (2006). *Introduction to Quantum Information Science.* Oxford University Press, Oxford.

Waddup, O., Blasiak, P., Yearsley, J.M., Wojciechowski, B.W., & Pothos, E.M. (2021). Sensitivity to context in human interactions. *Mathematics*, Vol. 9, No. 21, p. 2784; DOI: 10.3390/math9212784.

Wang, Z., Busemeyer, J.R., Atmanspacher, H., & Pothos, E.M. (2013). The potential of using quantum theory to build models of cognition. *Topics in Cognitive Science*, Vol. 5, No. 4, pp. 672–688; DOI: 10.1111/tops.12043.

Wojciechowski, B.W. (2015). Opiniowanie I psychologiczna analiza wyjaśnień, [in:] *Modele Psychologicznego Opiniodawstwa w Sprawach Karnych*, (ed.) Rode, D., GWP, Sopot, pp. 355–381.

Yearsley, J.M., & Busemeyer, J.R. (2016). Quantum cognition and decision theories: A tutorial. *Journal of Mathematical Psychology*, Vol. 74, pp. 99–116; DOI: 10.1016/j.jmp.2015.11.005

Yearsley, J.M., & Pothos, E.M. (2014). Challenging the classical notion of time in cognition: A quantum perspective. *Proceedings of the Royal Society B*, Vol. 281, No. 20133056; DOI: 10.1098/rspb.2013.3056.

Zhang, X., Qu, R., Chang, Z., Quan, Q., Gao, H., Li, F., & Zhang, P. (2022). A geometrical framework for quantum incompatibility resources. *AAPPS Bulletin*, Vol. 32, No. 17; DOI: 10.1007/s43673-022-00047-2.

INDEX

Note: **Bold** page numbers refer to tables; *italic* page numbers refer to figures.

For Product Safety Concerns and Information please contact our EU
representative GPSR@taylorandfrancis.com
Taylor & Francis Verlag GmbH, Kaufingerstraße 24, 80331 München, Germany

www.ingramcontent.com/pod-product-compliance
Ingram Content Group UK Ltd.
Pitfield, Milton Keynes, MK11 3LW, UK
UKHW020001200926
475702UK00015B/2416
* 9 7 8 1 0 3 2 5 1 4 8 2 6 *